SpringerBriefs in Applied Sciences and Technology

SpringerBriefs present concise summaries of cutting-edge research and practical applications across a wide spectrum of fields. Featuring compact volumes of 50–125 pages, the series covers a range of content from professional to academic.

Typical publications can be:

- A timely report of state-of-the art methods
- An introduction to or a manual for the application of mathematical or computer techniques
- A bridge between new research results, as published in journal articles
- A snapshot of a hot or emerging topic
- An in-depth case study
- A presentation of core concepts that students must understand in order to make independent contributions

SpringerBriefs are characterized by fast, global electronic dissemination, standard publishing contracts, standardized manuscript preparation and formatting guidelines, and expedited production schedules.

On the one hand, **SpringerBriefs in Applied Sciences and Technology** are devoted to the publication of fundamentals and applications within the different classical engineering disciplines as well as in interdisciplinary fields that recently emerged between these areas. On the other hand, as the boundary separating fundamental research and applied technology is more and more dissolving, this series is particularly open to trans-disciplinary topics between fundamental science and engineering.

Indexed by EI-Compendex, SCOPUS and Springerlink.

More information about this series at http://www.springer.com/series/8884

Paulo Roberto Bueno

Nanoscale Electrochemistry of Molecular Contacts

Paulo Roberto Bueno
São Paulo State University (UNESP)
Araraquara, São Paulo, Brazil

ISSN 2191-530X   ISSN 2191-5318   (electronic)
SpringerBriefs in Applied Sciences and Technology
ISBN 978-3-319-90486-3   ISBN 978-3-319-90487-0   (eBook)
https://doi.org/10.1007/978-3-319-90487-0

Library of Congress Control Number: 2018940398

This Springer imprint is published by the registered company Springer Nature Switzerland AG
The registered company address is: Gewerbestrasse 11, 6330 Cham, Switzerland

To exist is to change, to change is to mature, to mature is to go on creating oneself endlessly.

Henry Bergson

To my princess and our lovely little family

Preface

The idea for this short book emerged in 2016, when I was teaching a course on *Molecular Electronics and Electrochemistry*—the topic of this book—to doctoral students at the São Paulo State University (UNESP). At that time, I was unable to find a suitable book to guide my students on this topic. I was looking for an approach that offers a direct explanation of the basic fundamentals of quantum mechanics, aiming to solve nanoelectronics and nanoscale electrochemistry problems. The reason for this search was my clear awareness that, basically, electronics and electrochemistry are both somehow commonly grounded in quantum mechanics.

My initial intention was to write only lecture notes, but after teaching the course, it became clear that I had ended up with useful material that could be "easily" (little did I know!) converted into a concise introductory text for an intended audience moderately knowledgeable in the aforementioned interdisciplinary fields.

The merging of nanoelectronics and nanoscale electrochemistry can potentially modernise the way electronic devices are currently engineered or constructed. This book offers a conceptual discussion of this central topic, with particular focus on predicting the impact that uniting physical and chemical concepts at the nanoscale will have on the future development of electrochemical transistors and devices down to the molecular level. The electronic industry and its fabrication methods could be employed, for instance, to design efficient biosensors and supercapacitors.

Accordingly, I hope this book will be useful for physicists, chemists, material scientists and even biologists interested in electron transport and energy storage at the nanoscale. Previous knowledge in the fields of quantum mechanics, electrostatics, circuit analysis, electrochemistry and correlated disciplines would be desirable for readers aiming to gain an in-depth understanding of the concepts set forth herein. Also, a previous study or review of the scientific literature, focusing specifically on the electrochemical properties of electroactive monolayers, would be desirable and very useful.

This book was intended to be as short as possible, so it contains only three chapters. Chapter 1 introduces the concepts of chemical capacitance, relaxation resistance and quantum RC circuit dynamics and demonstrates how these concepts are translated into a language familiar to electrochemists. Chapter 2 demonstrates how chemical capacitance, which is the fundamental concept missing from the puzzle and that can unite electronics and electrochemistry, can be deduced from the first principles of quantum mechanics. The chapters are structured in such a way that Chap. 2 can be omitted in an initial reading, but an in-depth analysis would be impossible without an understanding of the deductions made in this chapter. Chapter 3 explains how the concepts introduced in the preceding chapters can be put to good use in different contexts, e.g. to describe applications involving supercapacitive and energy storage phenomena that are important in lithium-ion battery and supercapacitor devices, conductance in molecular wires, molecular electrochemical transistors and biosensors for molecular diagnostics.

Although the book is quite specific because of its inherent physical approach, it is at the same time fairly interdisciplinary. The text was intended to be comprehensive inasmuch as it introduces useful concepts for a broad range of current applications, although these concepts are not easy to grasp in a first reading. Knowledge about quantum transport and energy storage in molecular capacitors is often not readily assimilated, because these terms are unfamiliar to people involved exclusively with electrochemistry. Furthermore, such knowledge involves understanding quantum dynamics phenomena, which is inherently difficult at an undergraduate level. As for quantum mechanics, the fact is that there are only few individuals (myself not included) who understand its pervasive ambiguity; obviously, therefore, I do not expect to solve those difficulties.

All mechanical systems and living systems are subject to the temporal order of physics, where the event is more important than the entity *per se*. Nanoelectronics and nanoscale electrochemistry are only in their incipient stages of development and therefore destined to undergo continuous changes. Furthermore, as they are interdisciplinary fields, such changes will surely occur quite rapidly in the years ahead, particularly because the basic knowledge that underpins them is needed to drive the development of the next generation of computers and perhaps to greatly impact the development of molecular diagnostics technologies. Accordingly, I do not intend this book to be fully comprehensive or definitive, but simply to serve as an introduction and motivation to encourage other researchers to focus their attention on this fascinating topic.

Hence, I hope this book will serve as a provocative and a stimulating interdisciplinary exercise for those intending to labour on this promising scientific endeavour of bringing electronics and electrochemistry to a common ground and language.

Araraquara, Brazil   Paulo Roberto Bueno

Acknowledgements

The author gratefully acknowledges the Royal Society and São Paulo Research Foundation (FAPESP) for their financial support, which enabled investigations to be conducted in the author's laboratory in São Paulo, underpinning some of the information transmitted herein. This research served as an example that illustrates the application of some of the concepts introduced in this book. The author is also deeply grateful to Prof. Jason John Davis for his invaluable comments and discussions on the subject and for the numerous informal coffee and tea breaks we shared, as well as for inviting me to numerous dinners at the beautiful, magical and unforgettable hall of Christ Church College at the University of Oxford. Finally, I would also like to express my gratitude to him for hosting me at this university, where most of this book was written in 2017.

The kind assistance of members of my research group and of those students who provided raw data that served to illustrate some sections of this book must also be acknowledged, as well as their unabated enthusiasm for this subject. I also acknowledge Sandro Roberto Valentini, the Chancellor of São Paulo State University (UNESP), and my colleague Eduardo Maffud Cilli, Director of the Institute of Chemistry at UNESP, for their unceasing support for me and my academic career, which has been of inestimable value. Without their support, it is unlikely I would have been able to work as an academic visitor in the UK in 2017, during which I had the necessary peace and quiet to write this volume. Last but not least, I thank my wife for her unflagging patience, especially during the time I spent working on this book.

Oxford, UK
May 2018

Contents

About the Author

Prof. Paulo Roberto Bueno has a B.Sc. in Materials Science and Engineering, an M.B.A. in Economics and a Ph.D. in Theoretical Physical Chemistry. He is currently a Research Director at São Paulo State University and Head of the Physical Chemistry Department at that university. His main academic interest focuses on applications of electric and electrochemical spectroscopic methods, aiming to gain an in-depth understanding of the physical and chemical fundamentals of electron transfer and energy storage at the nanoscale. He has authored more than 170 papers, holds six licensed patents, and is one of the founders of Osler Diagnostics, a spin-off company from the University of Oxford, in the UK.

He has worked as a visiting academic and researcher on various occasions at several European universities, including the University of Paris and the University of Oxford. In the UK, some of his research projects have received awards from the Royal Society (including the Brian Mercer Feasibility and Newton Advanced Fellowship awards). He was endorsed as an exceptional talent in physical chemistry by the Royal Society and the UK government. Currently, he is a Research Fellow Director of the Royal Society, a Fellow of Royal Society of Chemistry and an invited member of the American Chemical Society. He is also a member of other scientific societies, among them the Electrochemical Society, the International Society of Electrochemistry and the Materials Research Society.

Keywords

Charge relaxation · Quantum capacitors · Nanoscale capacitors
Mesoscopic capacitors · Mesoscopic system · Quantum dynamic regime
Charge transfer resistance · Quantum resistive–capacitive circuit
Nanotechnology · Electrochemistry · Molecular electronics · Nanoelectronics
Supercapacitor · Batteries · Pseudocapacitance · Biosensors · Sensors

Introduction and Summary

Covalent attachment of molecules to electrodes and control of the properties of the junction represents the state of the art in molecular electronics and molecular electrochemistry. Molecular monolayers serve as building blocks for molecular electronics and have also been widely used in a variety of applications in electrochemistry. In this book, we demonstrate how the properties of these junctions are governed by the energy associated with the nanoscale capacitive characteristics of the molecular entities that form the junction and how they are related to electron transfer and transport. We demonstrate, in particular, the importance and the inherent electron dynamics that exists at the nanoscale and thus introduce the time-dependent electronic features that are accessed by impedance methods.

In other words, this book demonstrates that both the capacitive and resistive phenomena involved in electron dynamics throughout molecular-scale junctions, embedded in an electrolyte environment, are governed by mesoscopic principles and that appropriate mesoscopic physics is required to understand molecular junctions.

We demonstrate that a theoretical molecular ensemble (comprising individual quantum point contacts formed on an electron reservoir) is a particular case of a quantum resistance–capacitance circuit, which describes the electronic resonance between electrochemically accessible states and the electrode. Thus, electron dynamics operates in such a way that the electron transfer rate, which governs the rate of electrochemical reactions, is given by $k = G/C_{\bar{\mu}}$, where $C_{\bar{\mu}}$ is the electrochemical capacitance and G is the quantised conductance accompanying the Landauer formula, which describes the "quanta" of conductance. Surprisingly, this simple equation reconciles molecular electronics and electrochemistry.

The usefulness and significance of $k = G/C_{\bar{\mu}}$ are demonstrated in accessing the energy to charge molecular redox switches, in quantifying the discharge of these switches as the operative transducer signal in molecular diagnostics, also in determining the conductance of DNA nanowires and, lastly, in explaining the supercapacitance phenomenon of reduced graphene molecular layers and self-doped titanium oxide. In summary, this short book discusses the generalities associated with the $k = G/C_{\overline{\mu}}$ relationship, which is a time event and not a physical entity. The interpretation of $C_{\overline{\mu}} = G/k$, i.e. of electrochemical capacitance *per se*, which is associated with the energy needed for the event to occur, is introduced and deduced from the first principles of quantum mechanics, enabling us to grasp its general meaning.

Chapter 1
Introduction to Fundamental Concepts

This chapter discusses the importance of establishing a unified approach to electronics and electrochemistry. We explain how these disciplines are integrated when the circuit elements and their foundations are introduced from a nanoscale perspective, by modelling the system through a quantum dynamics circuit perspective rather than by using circuit elements defined in classical mechanics. Hence, the fundamentals of elementary physics and chemistry are explained, using terms such as chemical capacitance, electrochemical capacitance and relaxation resistance and how these circuit elements are interconnected to the timescale of the associated electron dynamics.

These are terms that govern the timescale of electron dynamics in electronics and electrochemistry at the molecular scale. As expected, "classical" or "traditional" electrochemistry is then shown to be simply an approximation derived from quantum mechanics.

1.1 The Importance of Nanoelectronics

Our understanding of the physics of nanoscale electronics considers it an urgent task that affects the future, as laid down in the roadmap for the semiconductor industry; gate widths for CMOS transistors today are smaller than 40 nm and are predicted to decrease between 10 and 5 nm [1] in the near future. The latter nanometric size [2–5] is close to the limit where quantum effects predominate that are not yet completely understood. For instance, designing an electronic device using molecules has been the ultimate goal in nanotechnology [6, 7], and molecular electronics has therefore been proposed as an alternative to silicon post-CMOS devices. The natural question [1] that arises in this scenario is: Will molecular quantum devices supplant traditional CMOS technology? The answer depends on the ability of researchers to understand, control and fabricate physical devices at such a diminutive scale.

The impedance of nanoscale devices such as coherent quantum resistance–capacitance (RC) circuits has been shown to violate Kirchhoff's law [8]. Therefore, the

© The Author(s) 2018
P. R. Bueno, *Nanoscale Electrochemistry of Molecular Contacts*, SpringerBriefs
in Applied Sciences and Technology, https://doi.org/10.1007/978-3-319-90487-0_1

predictability of their functioning and consequently the control of circuit devices in general, at this diminutive scale, are inherently challenging. Undeniably, the addition of capacitors and resistors to the classical approach does not hold true at the nanoscale. In particular, the rules governing the addition of resistances, in series [9, 10] or parallel [11], are qualitatively distinct from those of classical mechanics. Thus, electric circuits at the nanoscale are far more difficult to control, given that their electronic motion is wavelike, imposing fundamental limits on how small innumerable electronic devices can eventually be fabricated.

According to the above scenario, the resistance associated with charge relaxation differs from the usual quantum transport; therefore, the coherent quantum RC circuit does not comply with the Landauer formula [8, 12], although the incoherent one does [8]. Regardless of the charge transport regime (coherent or incoherent), quantum RC circuits are thus referential models for the dynamic operating regime of all quantum devices [13], although, surprisingly, this has been ignored in the fields of both molecular electronics and electrochemistry [7].

1.2 Molecular Electronics and Nanoscale Electrochemistry

Molecular electronics and nanoscale electrochemistry deal with electron movement through molecules via two different pathways [6, 7], i.e. electron transfer and electron transport. It should be noted that, strictly from a theoretical point of view, electron or charge transfer is a form of transport. As such, it is contained, for example, in a full approach to transport, including relaxation effects, using the Kadanoff–Baym formalism [14].

Nonetheless, in areas such as electrochemistry and molecular electronics, in practical terms, these phenomena (electron transfer and transport) can be differentiated by considering that electron transfer involves charge movement from one end of a molecule to the other, while electron transport pertains to electric current passing through a single molecule strung between electrodes, the latter constituting a two-terminal device configuration (see Fig. 1.1a), which differs from silicon-based electronics that rely critically on three-terminal devices (see Fig. 3.2a and traditional field-effect transistor architecture shown there).

Though highly desirable, the construction of gate terminals is a challenging task in molecular electronics [15], so break-junction experiments [16–18] (Fig. 1.1b) are used predominantly as two-terminal measurements. Even so, both two- and three-terminal configurations allow only limited mechanistic insights compared to the set-up required for time-dependent measurements (Fig. 1.1c) using redox-molecule-terminated wires as the probe.

Figure 1.1 illustrates a single molecule attached to electrodes as a basic component in molecular electronics [6] (Fig. 1.1a) and electrochemical break junctions [7, 19] (Fig. 1.1b), where a gold electrode is placed in contact with another electrode covered with sample molecules terminated with proper linkers. In Fig. 1.1, note that a third contact is possible and desirable (indicated as a gate), but this has been difficult

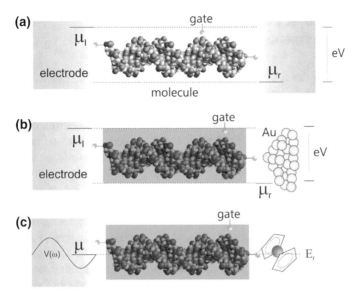

Fig. 1.1 Illustrations of a single molecule attached to electrodes as a basic component in: **a** molecular electronics [6], and **b** electrochemical break junctions [7, 19]. **c** Electrochemical gate configuration in which communication with the electrode occurs through molecular wires terminated with a redox probe [20]. *Reprinted (adapted) with permission from William W. C. Ribeiro, Luís M. Gonçalves, Susana Liébana, Maria I. Pividori, Paulo R. Bueno; Molecular Conductance of Double-stranded DNA Evaluated by Electrochemical Capacitance Spectroscopy, Nanoscale. Copyright (2016) Royal Society of Chemistry*

to achieve experimentally because it requires the gate to be placed close to the molecule for effective gate control [15]. Figure 1.1c shows an electrochemical gate configuration in which communication with the electrode is made through molecular wires terminated with redox probes (for an experimental example, see Chap. 3, Sect. 3.9) [20].

Regardless of the experimental set-up used to measure electron movement through molecules, the existing interpretation ignores quantum capacitive effects intrinsically associated with the electronic structure of the molecules which is perceivable by means of impedance spectroscopic methods [21,22]. Existing experimental configurations used in either molecular electronics direct conductance [6, 23] or charge transfer rate measurements [24] contain unnoticed information similar to that of quantum RC circuit dynamics.

1.3   Mesoscopic Physics and Time-Dependent Measurements

Mesoscopic physics is a collection of physical concepts that pertain to the properties of materials at the nanoscale, which is described as a scale between the molecular scale (the lower limit being the size of atoms) and a few hundred nanometres (nanoparticles or small proteins, for instance). The average properties of mesoscopic systems do not obey classical mechanics, but are affected by fluctuations around the average. Therefore, they greatly influence quantum mechanics effects, which must be treated at the quantum mechanics level without ignoring thermodynamics. In other words, a macroscopic device, when scaled down to a mesoscale size, starts revealing quantum mechanical properties that can no longer be ignored.

For example, at the macroscopic scale, the conductance of a wire increases continuously along with its diameter. However, at the mesoscopic scale, the wire's conductance is quantised and increases occur in discrete steps. Mesoscopic devices constructed, measured and observed experimentally during research must comply with a theoretical model in order to advance our understanding of the physics of such devices. Also, mesoscopic physics deals with the potential of building nanodevices using molecular assemblies; therefore, statistical thermodynamics is important. A frequent error of those not accustomed to quantum mechanics approaches and applications is to think classically, expecting that molecular entities have separate resistance or capacitance components that can be measured using traditional non-equilibrium methods. In fact, as in any quantum phenomenon, what is important is the event (which is time dependent) rather than the object (individual resistance or capacitance). For this reason, we will rely on time-dependent approaches rather than on direct current methods. Thus, the timescale of the process can be obtained by assuming the response of an average quantum resistive and capacitive (RC) circuit, which provides the quantum RC circuit response of a mesoscopic entity. This is the approach we will follow in this book. At the quantum or mesoscopic scale, resistance and capacitance can no longer be separated. From the experimental standpoint, an incorrect approach would be to ask what the resistance or capacitance of a molecule is. That is why, in this book, we ask what the timescale of the phenomena would be, and in so doing, demonstrate that what is important is the resistive–capacitive event itself and not the individual resistive and capacitive behaviours.

Indeed, time-dependent transport is an important and intriguing approach to reveal the physics underlying nanoscale man-made devices [8], yielding a wealth of information that could not be obtained otherwise. However, direct current (DC) measurements are quite limited. Time-dependent measurements based on impedance-derived spectroscopy set-ups [25, 26] have captured indispensable missing components of the puzzle involving molecular electrochemistry [20–22, 25, 27, 28]. Thus, electrical spectroscopy enables one to infer information about the coexistence of quantum capacitance and conductance phenomena, which determine the timescale of charge relaxation dynamics. Furthermore, impedance measurements can be taken based on either out of (biased) or in equilibrium (unbiased) potential difference, using an auxiliary electrode (known as a counter electrode) inserted in the electrolyte [22, 29]. Only one molecular contact is needed for impedance-derived measurements (see Fig. 1.1c), thus avoiding additional sources of interference inherently associated with the nature of the chemical bonds between the molecule and probing electrodes. Hence, impedance methods also reduce problems associated with the control of atomic-scale fine points of molecule–electrode contacts.

In Fig. 1.1, note that, unlike the classical DC transistor regime, the only way to access the properties of the channel (i.e. the accompanying resistance and capacitance) in a quantum RC regime is through time-dependent perturbation. Figure 1.1a and b show the differences between the left (μ_l) and right (μ_r) chemical potential of the electrons accompanying an external applied potential that drives electron movement, such that $d\mu = -edV$. In Fig. 1.1c, an equivalent relationship applies: $\bar{\mu} - E_r = -edV = d\bar{\mu}$ (see also Fig. 1.6a). The difference between $\bar{\mu}$ and the electrochemical potential of electrons in the accessible electrochemical states (E_r) is sustained by an electrochemical potential-driven concomitantly by the electrolyte and the polarisation of the working electrode with respect to an appropriate reference. This will be discussed in greater detail later in this book, but it is worth mentioning, briefly that the electrochemical potential difference built into the junction, $d\bar{\mu}$, is particularly accessible by measuring the electrochemical capacitance, $C_{\bar{\mu}}$, using sinusoidal potential perturbation methods [$V(\omega)$] such as impedance spectroscopy [21, 26, 30]. Details of the fundamentals of this capacitance will be given in Chap. 2 and applications in Chap. 3. The energy associated with the charge of $C_{\bar{\mu}}$ is directly proportional to $d\bar{\mu}$ $\left(e^2/C_{\bar{\mu}} = d\bar{\mu}/d\mathcal{N}\right)$ and reaches its maximum in equilibrium conditions, i.e. $\bar{\mu} = E_r$. Although differences between $d\mu$ and $d\bar{\mu}$ are implicit, they are discussed explicitly throughout this book.

The theoretical formulation of the problem is thus likely to determine the equilibrium electrostatic potential of a mesoscopic molecular structure (see Fig. 1.2a) modelled as a quantum point contact [31, 32], which is the fundamental building block to understand molecular electronics from a time-dependent perspective. Basically, a quantum point contact is a quantum dot [33–36] electronically coupled to an electron reservoir such as an electrode. Quantum point contacts (or quantum dots [33–36] in contact with electrodes) have been used as building blocks to outline mesoscopic transport [27, 28] in quantum devices comprising semiconductor nanostructures [37, 38].

To determine the equilibrium electrostatic potential of a quantum point contact, we must look beyond the conductive components whose states are of immediate interest. In a typical single molecular structure assembled on electrodes (Fig. 1.2a), the equilibrium electrostatic potential depends not only on the molecular conductor itself but also on the other neighbouring electrical charges provided by donors and acceptors, by gates and by chemical contacts (such as those solvating the conductor, redox accessible states, etc.; see Fig. 1.2b).

To find the equilibrium electrostatic potential, such additional neighbouring chemical species should be taken into account, but they have been critically disregarded in current theoretical approaches employed in both molecular electronics [14, 63–68] and electrochemistry [22, 29]. The existing gap between theoretical calculations

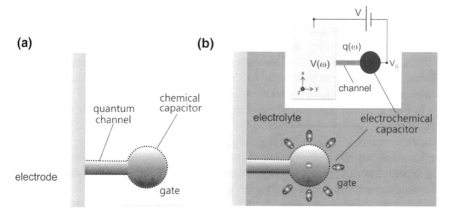

Fig. 1.2 a An individual quantum point contact structure comprising a nanoscale chemical capacitor, C_μ, (the series association of electrostatic and quantum capacitors) device assembled on an electrode and electronically coupled through quantum channels; **b** the same as in **a**, but within the point contacts immersed in an electrolyte environment where the associated potential of the gate is screened by the environment, illustrated herein by solvent molecules now surrounding the electrochemical capacitance, $C_{\bar\mu}$

and experimental data can be significantly reduced by understanding such over-looked aspects. The equilibrium electrostatic potential is of particular importance in time-dependent transport, since what is imperative is not the externally applied field (presumed to be controlled) but the total electric field generated by all the relevant charges, whether inside or away from the conductor, on a gate, or on the surface of the electrode.

Both theoretical and experimental analysis of time-dependent phenomena on a nanoscale is revisited herein, bearing in mind this important assignment. This analysis reveals that quantum RC dynamics are shielded by the electrolyte environment, impacting the use and interpretation of electron transfer theory in both molecular electronics and electrochemistry [22].

The implications are that the rate of electrochemical reactions is governed by $k = G/C_{\bar\mu}$, where $C_{\bar\mu}$ is the electrochemical capacitance and G is the quantised conductance accompanying the Landauer formula. $C_{\bar\mu}$ adequately incorporates the required equilibrium electrostatic potential intrinsically associated with the meaning of $C_{\bar\mu}$ (see below).

In Fig. 1.2, note that $C_{\bar\mu}$ is the series association of ionic [39] (double layer as a typical exemplar) and quantum capacitors [40, 41], thus constituting the missing physical structure of interest to interpret the time-dependent response of molecular electronics [14, 63–68] (in an electrolyte environment) and molecular electrochemistry [22, 29].

The electrical field screening in Fig. 1.2b is a critical aspect [28, 39] as will be discussed in further detail in Chap. 3, and is associated with the concept of $C_{\bar\mu}$ [29], in which the solvent chemical environment is crucial for electron dynamics to conform to electron transfer models [42, 43]. In the inset of Fig. 1.2b, note how this pictorial approach is consistent with perturbation theory, $C_{\bar{\mu}}(\omega) = dq(\omega)/dV(\omega)$, as detailed in Eq. (3.1). In the inset of Fig. 1.2b, $V = -\Delta\bar{\mu}/e$ conforms to the situation depicted in Fig. 1.1c.

As the text progresses and concepts become more familiar to readers, our ultimate goal is to illustrate how these concepts are involved in the astonishingly simple formulation $C_{\bar{\mu}} = G/k$, and how $C_{\bar{\mu}} = G/k$ applies to the design of molecular redox-switchable junctions [21, 41, 42] as nanoscale energy transducers and actuators for molecular diagnostic applications [44–48] and also how this formulation explains pseudocapacitive phenomena on the nanoscale, in general. However, the usefulness of the formula $C_{\bar{\mu}} = G/k$ introduced here is also demonstrated in calculating the quantised conductance in DNA acting [20] as nanowires and in explaining the intrinsic supercapacitance existing in reduced graphene layers [28] and self-doped titanium oxide [49].

1.4   Admittance, Conductance and Chemical Capacitive Ensembles at Zero-Temperature Approximation

Figure 1.2a corresponds to the representation of a macroscopic electronic reservoir (electrode) connected by an electronic channel consisting of quantum individual channels to a molecular discrete quantum capacitor. The latter corresponds to the existence of N conductor channels in which the transmission matrix is an $N \times N$-type matrix. Accordingly, it defines a Wigner-type [50] matrix, such that

$$\hat{N} = \frac{1}{2\pi j}S^{\dagger}\frac{\delta S}{\delta\bar{\mu}} \qquad (1.1)$$

From the unitary aspect of the S. matrix, $\hat{N}$ is assumed to be symmetrically positive with a spectrum τ_n/h, in which τ_n is the residence time of electrons in the n mode in the quantum structure of the mesoscopic capacitance. In Eq. (1.1), $j = \sqrt{-1}$ is the imaginary unitary number and h is the Planck constant. Following Eq. (1.1), the electronic density of states, $\mathcal{N}$ (DOS), can be rewritten as

$$\mathcal{N} = T_r\left[\hat{N}\right] = \frac{1}{h}\sum_n \tau_n. \qquad (1.2)$$

It can be now demonstrated that the admittance $G(\omega)$, at zero absolute temperature, is given by

$$G(\omega) = \sum_n \frac{1}{h/2e^2 + 1/\left(-j\omega C_{q,n}\right)}, \qquad (1.3)$$

where $C_{q,n} = e^2\tau_n/h$ is the quantum capacitance (the meaning of quantum capacitance will be clarified later) associated with each of the channels and e is the elementary charge of the electron.

Equation (1.3) is a model that also applies to a molecular assembly of quantum channels and capacitors attached to a macroscopic electrode, as illustrated in Fig. 1.6. Each channel provides an equivalent dissipative contribution of $h/2e^2$ and a reactance contribution equivalent to the quantum capacitance $C_{q,n}$ associated with the n mode of the electronic channel. The quantum admittances of the different modes add up, translating to parallel branches of the quantum impedances, such that $h/2e^2 + 1/(-j\omega C_{q,n})$, which, in turn, translates to a series equivalent circuit of R_q and C_μ, i.e.

$$R_q = \frac{h}{2e^2} \frac{\sum_n \tau_n^2}{\left(\sum_n \tau_n\right)^2} \tag{1.4}$$

and

$$C_\mu = \frac{C_{ext}\left(\frac{e^2}{h}\sum_n \tau_n\right)}{C_{ext} + \left(\frac{e^2}{h}\sum_n \tau_n\right)}, \tag{1.5}$$

where C_μ is now the chemical capacitance and C_{ext} is the external capacitance associated with the external potential imposed on the system (see Fig. 1.6).

The meaning of C_μ and C_{ext} will be given in further detail in this book, in specific sections reserved for this purpose. By now, it should be clear that Eqs. (1.4) and (1.5) are deduced based on scattering theory (known since the 1990s [13]) and are deduced herein in the same way as the Landauer formula, derived for direct electric currents (DC).

1.5 Admittance, Conductance and Chemical Capacitive Ensembles at Finite Temperature

At finite temperature, the electron distribution follows $\langle \mathcal{N} \rangle = (1 + \exp[-eV/k_BT])^{-1}$, in which the energy width is given by k_BT, from where the quantum admittance, $G(\omega)$, translates into

$$G(\omega) = \int d\mu \left(\frac{d\langle\mathcal{N}\rangle}{d\mu}\right)_{T,\tilde{v}} \frac{1}{h/2e^2 + \left[-j\omega C_\mu(\mu)\right]} \tag{1.6}$$

where $C_\mu = e^2[d\mathcal{N}(\mu)/d\mu]$.

The system is thus equivalent to a multichannel system in which the N conducting channels are replaced by a continuous infinite number of parallel channels weighted by the Fermi–Dirac distribution. As before, the admittance of the RC circuit is obtained by means of a conduction channel at a given temperature, considering the admittance $G(\omega)$ in series with the electrostatic capacitance C_{ext}, such that

$$G(\omega) = \frac{-j\omega C_\mu}{1 - j\omega R_q C_\mu} \tag{1.7}$$

where

$$R_q = \frac{h}{2e^2} \frac{\int \left(-\frac{\mathrm{d}\langle \mathcal{N}\rangle}{\mathrm{d}\bar{\mu}}\right) \mathcal{N}(\bar{\mu})^2 \mathrm{d}\bar{\mu}}{\left[\int \left(-\frac{\mathrm{d}\langle \mathcal{N}\rangle}{\mathrm{d}\bar{\mu}}\right) \mathcal{N}(\bar{\mu}) \mathrm{d}\bar{\mu}\right]^2} \tag{1.8}$$

and

$$C_\mu = e^2 \int \left(-\frac{\mathrm{d}\langle \mathcal{N}\rangle}{\mathrm{d}\bar{\mu}}\right) \mathcal{N}(\bar{\mu}) \mathrm{d}\bar{\mu} \tag{1.9}$$

and where, lastly,

$$C_\mu = \frac{C_{ext} C_q}{C_{ext} + C_q} \tag{1.10}$$

Note that when embedded in an electrolyte (see Fig. 1.2b), C_{ext} translates into C_i, which is the capacitance associated with an electrical field screened by an electrolyte environment [see also Eq. (1.14), where C_i replaces C_{ext}].

As will be explained, instead of using the term C_μ, we will now refer to $C_{\bar{\mu}}$. In fact, the differences between C_μ and $C_{\bar{\mu}}$ will be clarified later and will thus serve to demonstrate that nanoelectronics, and electrochemistry are conceptually equivalent and that the differences pertain only to how the electrical field is screened by the environment in one or the other case. Particularly, it will be demonstrated that, in electrochemistry, electric field screening is governed by the properties of the electrolyte.

Now, consider $\hbar\ell \sim D\Delta/2$ as the characteristic width of the discrete peaks of electronic density of states, keeping in mind that D is the electron transmittance, while Δ is the difference between the peaks (homogeneous dispersion has been assumed). According to the latter inference, then, it is possible to predict the existence of a quantum capacitance peak whenever the Fermi level of the reservoir is at a level aligned with that of the mesoscopic capacitance or an ensemble of mesoscopic capacitors. Thus, to calculate the admittance of the mesoscopic RC circuit associated with R_q and C_μ, the displacement current between the mesoscopic capacitance and the gate carrying the excitation must be taken into account.

Accordingly, let us now consider a finite temperature situation where the conduction, $\hbar\ell \ll k_B T$ (or $k_B T \gg \Delta$), will involve several levels of energy, so that it can be demonstrated that R_q is given by

$$R_q \sim \frac{h}{e^2} \frac{1}{D},$$
(1.11)

which corresponds to the Landauer resistance between two reservoirs, i.e. the situation depicted in Fig. 1.1a, whereas no hypothesis has been proposed about the loss of coherence in the mesoscopic capacity. Thus, thermal interference leads to the disappearance of coherence, allowing the characteristic time width of an electronic wave packet to be defined. The latter is then given by $h/k_B T$, while the time it takes for an electron exchange, i.e. its emittance from the reservoir to the capacitive centres and its return, is h/Δ.

In summary, when $k_B T \gg \Delta$, the electron cannot interfere with itself and the quantum bodywork of the mesoscopic capacitance behaves like a reservoir. Particularly, in the case of a two-dimensional molecular ensemble containing electrochemical accessible states, it behaves like a two-dimensional electron gas [41]. The next section describes how this theoretical framework combines with electrochemistry.

1.6   State of the Art of Nanoscale Electrochemistry

Electrochemistry governs the properties of important energy and sensor devices such as photoelectrochemical solar cells [51, 52], fuel cells [53–55], lithium-ion batteries [56–58], and molecular recognition devices [59], and it is equally important in biochemistry [60, 61], dominating the processes associated with cell respiration [62, 63], photosynthesis [64–66], etc. Electronic and ionic dynamics govern the timescale properties of these devices and biochemistry processes in general, but a critical question that arises, among others, is "How do electron dynamics operate at the molecular level [67–74] in a complex solvent environment?" [75, 76].

This question is important in that intrinsically biological electrical processes (such as synapses [77, 78], which, in essence, involve an electrochemical occurrence) are slower than any known electronic signal, although this has apparently not been a problem for efficient energy management, storage, and conversion, as observed in living systems [55, 79]. The impact of the electrolyte environment on the nanoscale system, by attenuating the electron dynamics, results in a larger timescale of the charging events, which is an inevitable price to be paid. Accordingly, bioelectronics (or bio-electrochemistry) is fated to be slower than electronics, albeit not necessarily less efficient. Evidence in favour of the latter argument can be obtained by comparing the brain's synapses to a computer's processing capability.

Consider ionic intercalation processes such as those that occur in lithium-ion batteries, for instance, [53, 80, 81]. It is well known that both electronic and ionic densities increase by the same amount upon ionic interaction with anode and cathode electrodes, and hence, electronic conductivity usually exceeds ionic conductivity (although there are exceptions [51, 82–84] which must be treated separately). The operational timescale of the device is therefore determined by the slowest ionic charge carrier, but would this also be true at the molecular scale? In the absence of diffusion [42, 85, 86] (see below), what would the dynamics be?

To illustrate the above, consider a hypothetical Li_xMO_y solid compound, where x is the lithium-ion content that can be modified in a metal oxide compound MO_y (serving as working electrode). The modification of the Li_xMO_y composition determines the working electrode potential, such as $eV = -[\mu(x) - \mu_r] = -\Delta\mu$, where the chemical potential variation $\mu(x)$ is determined with respect to a chemical potential of the reference μ_r. Ignoring the effect of the electrolyte external potential, the measured chemical capacitance (per unit of volume) is given by $C_\mu = e^2\mathcal{N}[dx/d\mu(x)]$, which can then be commonly determined as the inverse derivative of the equilibrium–composition curve [87, 88]. $\mathcal{N}$ is the electron occupancy in the solid compound that attends ionic intercalation (the overall electric neutrality is maintained). It is thus clear that the ionic component dominates the dynamics operating in this hypothetical electrochemical compound, since the slowest transport the ionic one. The latter compound, albeit hypothetical, quite accurately exemplifies how cathodes and anodes operate in lithium-ion batteries [51].

Nonetheless, in electronic conducting nanoscale polymers and redox molecular switch devices, the latter generally consisting of a redox molecular monolayer, the energy landscape of electronic states dictates the electrochemical response in such a way that $C_{\bar{\mu}} = e^2[d\mathcal{N}/d\bar{\mu}]$. In this formulation, ionic diffusion is absent and electrochemical reversibility is high; moreover, the effect of the electrolyte is now taken into account, so $C_{\bar{\mu}}$ and $\bar{\mu}$ are used in place of C_μ and μ. Furthermore, in the particular case of redox switches, due to the intrinsic molecular scale of the components bridging the redox centres to the electrode (see further experimental examples in Chap. 3), the electron transfer or transport is driven on a size scale of 1–3 nm, which is close to the minimum predicted width of the CMOS transistor gate. Hence, at this diminutive scale, there are quantum electrochemical RC characteristics that enable one to identify a rich variety of physical phenomena inherently associated with the electronic structures of molecular components [41].

In the case of molecular components that have accessible electronic or electrochemical (redox) states, their physical phenomena are accompanied by intrinsic quantum RC features. Accordingly, electrochemistry itself can be rationalised by considering the quantum electronic RC circuit a useful physical reference, which will herein be referred to as the quantum electrochemical RC circuit because of the attenuation of the electron dynamics caused by the solvent (see Fig. 1.2b), which is characteristic of electrochemistry.

1.7 Consequences of the Absence of Ionic Diffusion in Electrochemical Processes at the Nanoscale

Electroactive species adsorbed in electrodes (such as redox molecular switches) generally present a highly reversible electrochemical behaviour. This is due to the absence of mass transport limitations, such as diffusion processes, which are involved in the displacement of ions from the bulk of solution to the surface of the electrode where the electron charge transfer actually occurs. An example is given in Fig. 1.3a, which shows an electric current versus potential cyclic pattern, known as a cyclic voltammogram (CV). This current–voltage pattern corresponds to the response of a redox molecular monolayer assembled on a gold electrode. The molecular layer is an ensemble of electroactive molecules self-assembled on an electrode composed of 11-ferrocenyl-1-undecanethiol units.

The plot illustrates the dependence of the current density on the scan rate, s. The inset shows how the oxidation and reduced peak potential vary linearly as a function of the scan rate, as predicted based on the Laviron approach [85, 86]. Figure 1.3b shows that the response is in agreement with Eq. (1.12). It demonstrates that a constant value $(C_{\bar{\mu}} = e^2\Gamma/4k_BT)$ is obtained when the current (or the current density) is normalised by the scan rate. Accordingly, normalised CVs collapse into a master curve. This is particularly remarkable at the Fermi level (or formal potential) of the electrode, confirming that quantum dynamics RC theory is needed, which will be introduced in this chapter. This theory adequately encompasses DC behaviour and traditional electrochemistry.

In short, one can see that the diffusionless process occurs because the redox species is strongly adsorbed (covalent attachment, for instance, generally results in highly reversible electrochemical processes) on the electrode [42]. In an electrochemical process involving the transfer of a single electron per molecule, considering a diffusionless situation, the Faradaic electric current density is directly proportional to the scan rate ($s = \mathrm{d}V/\mathrm{d}t$), such that [42]

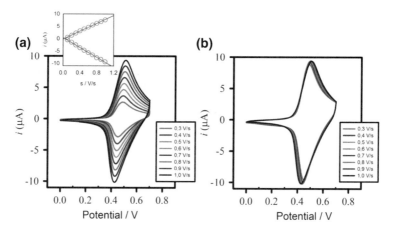

Fig. 1.3 a Cyclic voltammograms of a molecular layer composed of 11-ferrocenyl-1-undecanethiol, showing the dependence of current density on the scan rate, s. The inset shows how the oxidation and reduced peak potential vary linearly as a function of scan rate, as predicted by the Laviron approach [85, 86]. **b** As expected from Eq. (1.12), when the current (or current density) is normalised by the scan rate, one obtains a constant value $\left(C_{\bar{\mu}} = e^2\Gamma/4k_BT\right)$

$$j = \frac{F^2\Gamma}{4RT}s = \frac{e^2\Gamma}{4k_BT}s \qquad (1.12)$$

where F is the Faraday constant, R is the constant of the ideal gas, T is the absolute temperature, and Γ is the molecular coverage of the electrode (the amount of covalently adsorbed molecules per unit of area). Note that $F/R = e/k_B$, where e is the elementary charge of the electron and k_B is the Boltzmann constant.

When the scan rate-dependent electron transfer rate, k_s, is related to the amount of Faradaic charge, such that $k_s(\eta) = j/q$, where q is the total charge, it can be varied as a function of overpotential displacement, $\eta = V_p - V_r$, where V_p is the potential of the peak (see Fig. 1.3) and V_r is the formal potential (in terms of energy, and this is related to the Fermi level of the redox process, E_r). Using $k_s(\eta) = j/q$, the rate constant for electron transfer at a particular overpotential can be determined to obtain a plot of η versus $\log[s(\eta)]$. The latter is known as a Tafel plot [89], from which k can finally be determined.

As an alternative to Laviron analysis, Eq. (1.12) can be equivalently resolved by considering a general definition of electrochemical charge associated with the electrochemical capacitance, such that $dq = C_{\bar{\mu}}dV$. The variation of charge as a function of time provides [41] the electrical current associated with the charge of the electrochemical capacitance, such that $j = (dq/dt) = C_{\bar{\mu}}(dV/dt) = C_{\bar{\mu}}s$. Upon comparison the latter expression with Eq. (1.12), one can see that $C_{\bar{\mu}} = e^2\Gamma/4k_BT$. In turn, $C_{\bar{\mu}} = e^2\Gamma/4k_BT$ is equivalent to Eqs. (3.9) and (3.10), as discussed in Chap. 3, and is obtained based on quantum mechanics reasoning combined with an analysis based on statistical thermodynamics. At the Fermi level of the electrochemical reaction, $\langle \mathcal{N}\rangle(1 - \langle \mathcal{N}\rangle) = 1/4$, *following statistical thermodynamics requirements* (further details will be given in Chap. 3).

Therefore, the quantum electrochemical RC circuit encompasses classical electrochemistry, though the physical interpretation is obviously absolutely different and based essentially on quantum mechanics rather than on classical circuit elements or traditional electrochemistry.

Accordingly, let us now reiterate that Fig. 1.2 illustrates single-contact nanoscale electrochemical junctions containing chemical (C_μ) and electrochemical $(C_{\bar{\mu}})$ capacitive gates connected to an electrode. These capacitive models are introduced with (Fig. 1.2b) and without (Fig. 1.2a) the presence of an electrolyte that governs equilibrium potential differences. The case depicted in Fig. 1.2a is currently used in modelling two-dimensional density of states contained in solid-state devices [40, 90, 91]. An electrolytic environment (Fig. 1.2b) generates additional issues and must be treated accordingly, using $C_{\bar{\mu}}$ instead of C_μ (see previous justifications). The former capacitance is the physical picture that conforms to electrochemical junctions. Electrochemical capacitive gates are spatially separated from an electrode by quantum channels [92], which, pictorially, account for the path electrons must follow during physical events such as electron transfer or transport.

When the tiny capacitive gates are embedded in an electrolyte, the timescale of electron dynamics (usually at gigahertz frequencies for nanoscale electronic circuits) is attenuated (to ranges between kilohertz and millihertz) by the electrolytic environment, in a scheme that adheres to electron transfer rate theory [21, 22]. Therefore, the distinction between C_μ and $C_{\bar{\mu}}$ must be observed carefully by highlighting (see below) the electric field screening for each particular situation.

Additionally, it will be shown that the electrochemical control of a quantum capacitive electrochemical contact (in contrast to pure electrostatics) is the component that allows us to assess this previously unknown knowledge about electron transfer phenomena, bringing together electronics and electrochemistry [43, 93–95].

The next section demonstrates how electronics and electrochemistry are integrated. This will be done by introducing concepts associated with nanoelectronics theory [2, 5, 96] and then explaining how it conforms to electrochemistry.

1.8 Chemical Capacitance and Charge Relaxation

Figure 1.4 depicts charge separation due to a potential difference, V, between two plate electrodes (Fig. 1.4a) and between molecular donor (D) and acceptor (A) states (Fig. 1.4b). This figure clearly demonstrates that these charging processes differ and, in the case of molecular charge donor and acceptor states, how they depend essentially on the electronic structure of the molecular entities.

Therefore, Fig. 1.4a illustrates the case in which only electrostatic contributions are present; i.e., in charging metallic plate electrodes, the resulting capacitance is C_e (the rationale for this is discussed later herein). On the other hand, Fig. 1.4b depicts a case in which molecular structures are charged (where quantum effects are non-

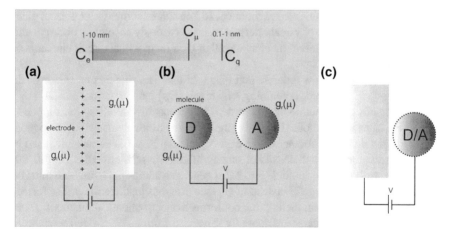

Fig. 1.4 a Separation of charges (due to a potential difference, V) between two plate electrodes. **b** Separation of charges between molecular donor (D) and acceptor (A) states. **c** Situation in which the quantum capacitance (be it of a D or an A state) can be approximated to $C_q = e^2 \mathcal{N}$ at zero absolute temperature (non-thermalised condition) and to $C_q = e^2 g_r(\mu)$ at a finite temperature [41]

negligible), and therefore, there are quantum capacitive contributions contained in the definition of C_μ, which is a universal chemical capacitance (existing in series) that considers both electrostatic and quantum contributions.

The quantum capacitive component of C_μ quantitatively encompasses HOMO and LUMO contributions when the charge moves into atomic or molecular structures (see Chap. 2 for additional details on this) [41]. For real problems, a macroscopic probe is needed to charge molecular components; therefore, Fig. 1.4c illustrates a situation in which the quantum capacitance (be it of D or A) can be approximated to $C_q = e^2 \mathcal{N}$ in the non-thermalised condition or to $C_q = e^2 g_r(\mu)$ for finite temperatures [41]. This picture is essentially the same as the one that models quantum dots [32–35] electronically coupled to electrodes [41].

Since we now have a better pictorial view of how charging processes compare in different scales and situations (Fig. 1.4), we can combine the results obtained from Eqs. (1.4) and (1.5) to describe the response of an ideally quantum RC electronic circuit at the zero-temperature limit. This provides the timescale to charge a single quantum capacitive state from an electrode (Fig. 1.2a), which is [8, 13, 97]

$$\tau = \left(\sum \frac{h}{2e^2 D} \right) \left(\frac{1}{C_e} + \frac{1}{C_q} \right)^{-1} = \left(\sum \frac{h}{2e^2 D} \right) \left(\frac{1}{C_\mu} \right)^{-1} \qquad (1.13)$$

where C_e is the electrostatic capacitance and $C_q = e^2 \mathcal{N}$ is the quantum capacitance.

As is implicit in Eq. (1.13), it can be noted that C_μ is the series association of electrostatic capacitance, which plays the role of external capacitance, and C_q. $h/2e^2 D$ is the Landauer formula for a single quantum channel, in which, as mentioned earlier, D is the electron transmission of an individual point contact. In an ideal point contact, D equates with a unit, so that $h/2e^2$ is a universal constant (as previously discussed in Sects. 1.4 and 1.5). The sum of the resistance of different quantum channels defines the relaxation resistance as $R_q = \left(\sum h/2e^2 D \right)$.

Effectively, the quantum of resistance or relaxation resistance is associated with the probability of electron transmission of the quantum channel, as illustrated in Fig. 1.2, which separates the accessible electronic state in C_μ from the electronic reservoir, which, in turn, is the electrode illustrated in Fig. 1.2. Thus, Eq. (1.13) summarises the properties of a coherent quantum RC electronic circuit (at gigahertz frequencies), which describes the electron dynamics of this ideal capacitive point contact.

The meaning of C_μ is universal, as exemplified in Fig. 1.4; classical and quantum capacitances are two limits of fictitious capacitive metrics (depicted at the top of Fig. 1.4). According to classical electromagnetism, an electrostatic capacitance is dependent only on the geometric factors of the device [41, 98]. By taking the plate capacitor as an exemplar, the electrostatic capacitance per unit of area is calculated as $C_e = \varepsilon_r \varepsilon_0 (1/L)$, where L is the distance between the plates of the capacitor (Fig. 1.4a), ε_r is the relative dielectric constant of the environment, and ε_0 is the dielectric vacuum permittivity.

On the other hand, quantum capacitance is defined on the atomic scale as $1/C_q = 1/e^2[1/g_l(\mu) + 1/g_r(\mu)]$ [13, 25], in which $g_l(\mu)$ and $g_r(\mu)$ are the left- and right-hand electron density of states of two separate mesoscopic entities (atoms or molecules referred to as D and A entities), as illustrated in Fig. 1.4b. The quantum capacitance [13, 25, 29] is therefore dependent on the chemical potential differences of the electrons [41, 43] of left and right molecular entities (donor and acceptor).

Realistically, assuming an intermediate mesoscopic situation between this classical limit and the quantum limits, we defined the chemical capacitance as a series association of C_e and C_q; $1/C_\mu = 1/C_e + 1/C_q$ (Fig. 1.4b). As expected for the classical limit, $g_l(\mu)$ and $g_r(\mu)$ are numerically large densities of states associated with the left-hand side and right-hand side metallic contacts, providing the limit in which C_μ equates to C_e and quantum capacitive effects are depreciable (Fig. 1.4a). $C_q = e^2 \mathcal{N}$ is a possible and in fact quite a good approximation (not demonstrated) of $1/C_q = 1/e^2[1/g_l(\mu) + 1/g_r(\mu)]$, employed for circumstances in which there is a difference of potential between a quantum point contact and an electrode, as depicted in Fig. 1.4c.

In summary, small atomic or molecular devices (as a single contact or a collection of quantum point contacts) attached to an electrode in a vacuum or ambient air can be experimentally studied by the linear response of sinusoidal perturbations provided by impedance spectroscopy methods. Because the molecules attached to an electrode are adequately modelled as quantum point contacts (individually or as ensembles), the impedance spectroscopic response is predicted to follow Eq. (1.13).

1.9 Electrochemical Capacitance and Electron Transfer Rate

An equivalent nanoscale electrochemical RC circuit [as stated in Eq. (1.13)] exists when the hypothetical quantum point contact is embedded in a dielectric continuum (Fig. 1.2b) or electrolyte, so that C_μ is now replaced by $C_{\bar{\mu}}$ (the electrochemical capacitance), and hence [41].

$$\frac{1}{C_{\bar{\mu}}} = \left(\frac{1}{C_i} + \frac{1}{C_q} \right) \qquad (1.14)$$

where C_i is the ionic capacitance of the interface [39, 99], which is normally modelled by a double-layer phenomenon. A difference should be highlighted between this depiction of the charging events and the one in the preceding section, simply because C_i differs from C_e, as illustrated in Fig. 1.5. The ionic capacitance per unit of area is $C_i = \varepsilon_r \varepsilon_0 \kappa$, where κ is dependent on ionic strength; κ is the inverse of the Debye length [39] and so differs from the geometric factor $1/L$ contained in $C_e = \varepsilon_r \varepsilon_0 (1/L)$.

For instance, in the absence of Faradaic activity, $1/C_q$ is depreciable since the electronic density of states $\mathcal{N}$ is inaccessible for charging by a potential difference

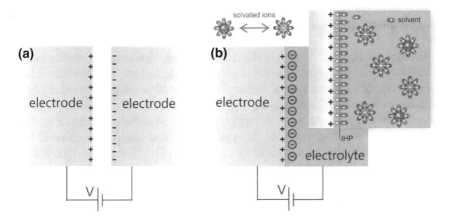

Fig. 1.5 **a** Electrostatic plate capacitor in which the capacitance per unit of area is inversely dependent on the distance between the plates, i.e. $1/L$. **b** When an electrode plate is in contact with an electrolyte, in the absence of electroactive species, an analogous electrochemical plate capacitance exists, but ionic effects must be taken into account. *Reprinted (adapted) with permission from Joshua Lehr, Justin R. Weeks, Adriano Santos, Gustavo T. Feliciano, Melany, I. G. Nicholson, Jason J. Davis and Paulo R. Bueno; Mapping the ionic fingerprints of molecular monolayers, Physical Chemistry and Chemical Physics. Copyright (2017) Royal Society of Chemistry*

established with the electrode, corresponding to a situation where polarisation occurs without charge transfer (referred to as a non-Faradaic event) [39]. Equation (1.14) predicts a double-layer capacitive model and also includes other capacitive effects associated with the charging of nanoscale molecular ensembles. Whether or not accessible electronic states exist, there is an innate presence of quantised capacitive effects usually embedded within what is described as "pseudocapacitance" [22].

Under such circumstances, then, both ionic and electronic contributions must be considered in a combined electrochemical capacitance $C_{\bar{\mu}}$, which includes a consideration of Thomas–Fermi electrical field screening [$\left(1/C_{\bar{\mu}} = 1/C_i + 1/C_q\right)$ [28, 39] to which ionic (C_i) and quantic $\left(C_q\right)$ terms contribute] and $\kappa = \left(C_{\bar{\mu}}/\varepsilon_r\varepsilon_0\right)^{1/2}$, which is the Thomas–Fermi wave vector [28, 39]. The meaning of κ is therefore important, since it is dependent on $C_{\bar{\mu}}$ and on the dielectric environment, $\varepsilon_r\varepsilon_0$.

Both ionic (distinguishable particles) and electron (indistinguishable particles) are embedded in the experimentally resolved $C_{\bar{\mu}}$ of an electrochemical interface. If a Boltzmann approximation is assumed, in a pure non-Faradaic charge of the interface, then $\mathcal{N} \propto \exp[-eV/k_BT]$ (k_B is the Boltzmann constant and T is the absolute temperature), where the fluctuations in the electric potential follow classical behaviour.

On the other hand, assuming a pure Faradaic charge of the interface, considering only indistinguishable particles, then $\mathcal{N} \propto (1 + \exp[-eV/k_BT])^{-1}$ and the fluctuation of electric potential is distributed according to Fermi–Dirac statistics. Note that if the ionic contribution predominates, then $\kappa = \left[\left(2e^2\mathcal{N}\right)/(\varepsilon_0\varepsilon_rk_BT)\right]^{1/2}$ is recovered from $\kappa = \left(C_{\bar{\mu}}/\varepsilon_r\varepsilon_0\right)^{1/2}$, in accordance with the Debye–Hückel model [100, 101], demonstrating that double-layer phenomena are only an approximation of Eq. (1.14) [39].

In other words, as can be seen in Fig. 1.5, these capacitances are dependent on the inverse of the Debye length, κ, which is modulated by the ionic strength resembling the electrostatic capacitor model, although, in terms of the atomic structure of the junction, it is manifold and the analogy must be drawn with restrictions. These capacitances can be modelled by the simplest capacitive model, known as a double-layer structure, which is negatively charged as illustrated in Fig. 1.5b, in which the structure closest to the electrode is known as the inner Helmholtz plane (IHP) of the double-layer model [39, 89].

As expected, double-layer phenomena prevail in the absence of accessible electronic states of molecular layers assembled on the electrode (see examples in Chap. 3, Sect. 3.6). This is also the case of graphene oxide, in contrast to reduced graphene oxide (see, also, a further discussion about supercapacitance phenomena in graphene, Fig. 3.11), where, henceforth, Debye rules over Thomas–Fermi screening. On the other hand, in situations where there is a significant accessible nanoscale electron density of states, $C_{\bar{\mu}}$ is dominated by C_q and the electrical field screening is dominated, accordingly, by electron dynamics attenuated by the electrochemical environment [21, 29, 39], and hence governed by Thomas–Fermi rather than Debye electric field screening.

Now let us demonstrate what the timescale of an electrochemical process would be, following Eq. (1.13), but whose capacitance is $C_{\bar{\mu}}$. The latter is the case where the charge of the mesoscopic capacitance is embedded in an electrolytic environment [22]. This obviously corresponds to the timescale of the electrochemical reaction conforming to a single electron transfer, such that [22, 43].

$$k = \left(\sum \frac{h}{2e^2 D}\right)^{-1}\left(\frac{1}{C_i} + \frac{1}{C_q}\right) = \left(\sum \frac{h}{2e^2 D}\right)^{-1}\frac{1}{C_{\bar{\mu}}} \qquad (1.15)$$

where k is well known as the electron transfer rate constant. Equation (1.15), which was validated previously, represents the quantum RC circuit model for electrochemistry, hypothetically defined at zero-temperature approximation for a single molecule embedded in an electrode.

Note now that Eq. (1.15) can be rewritten in the form of $C_{\bar{\mu}} = G/k$, where $G = 1/R_q$ is a quantised conductance and R_q is the quantum limit of a charge transfer resistance (R_{ct}), as referred to in electrochemistry textbooks [90].

According to Sect. 1.5, it is evident that $C_{\bar{\mu}} = G/k$ is equivalent to the expected response of a set of molecules assembled on an electrode, as depicted in Fig. 1.6 [22, 43]. It should be pointed out that Eq. (1.15) provides a generally valid description of mesoscopic systems and does not require a perfectly conducting channel with a quantised unit transmission. Thus, in this case, there is a relationship between $\sum 2e^2 D/h$ and k that can be generalised as a relationship between G and k in distinct situations.

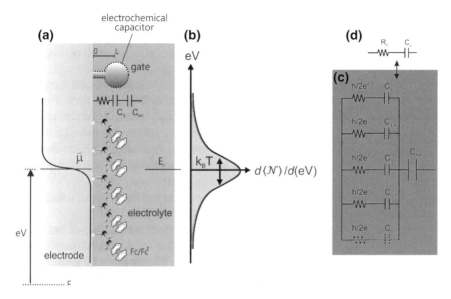

Fig. 1.6 **a** Molecules assembled on an electrode, forming a molecular film embedded in an electrolyte. **b** Expected thermal broadening given by $d\langle\mathcal{N}\rangle/d(eV) \propto \langle\mathcal{N}\rangle(1 - \langle\mathcal{N}\rangle)$ (see further details in Chap. 3). This is predicted as Gaussian-shaped thermal broadening (indeed, it has been observed experimentally, as illustrated in Fig. 3.8b in Chap. 3). **c** Quantum electrical circuit representation (red box) of the molecular films shown in (a). **d** Equivalent circuit of the ensemble depicted as series R_q and $C_{\bar{\mu}}$ circuit elements. L corresponds to the thickness of the electron path through the molecular layer

The response that confirms the relationship between $C_{\bar{\mu}}$ and G, i.e. $C_{\bar{\mu}} = k/G$, is shown in Fig. 1.6 for alkane molecular films [22]. Also, interesting is the correspondence of $C_{\bar{\mu}}$ with first-principle density functional theory (DFT) calculations [29, 41], which will be developed in detail in Chap. 2. The latter relationship demonstrates the generality of the theoretical framework developed here. The connection of Eq. (1.14) or $C_{\bar{\mu}}$ with first principles of quantum mechanics is important because it enables us to understand electrochemistry from the conceptual standpoint of quantum chemistry [102]. Therefore, theoretical analysis relies not only on classical mechanics, which has been preventing electrochemists from advancing their understanding of how molecules attach to electrodes, but also allows us to demonstrate that molecular electronics and electrochemistry are conceptually integrated.

An immediate interpretation of Eq. (1.15) is that it connects the concepts of electron transfer rate and quantum of conductance in such a way that electron transfer and transport are totally interchangeable through the microscopic accessing of $C_{\bar{\mu}}$.

For instance, Fig. 1.6a illustrates the case where molecules assembled on an electrode surface form a molecular film embedded in an electrolyte. This particular situation can be modelled by an ensemble of individual quantum point contacts. If the thermal energy is considered, as shown in Fig. 1.6b, it is possible that the expected thermal broadening given by $d\langle\mathcal{N}\rangle/d(eV) \propto \langle\mathcal{N}\rangle(1 - \langle\mathcal{N}\rangle)$ plays an important role (this will be discussed and demonstrated in Chap. 3, Sect. 3). This predicted Gaussian-shaped thermal broadening is experimentally measured, [21, 27] as shown in Fig. 3.8b, and k spreads [103] as a consequence of changes in C_{ext}, which consequently affects $C_{\bar{\mu}}$ according to [21, 29, 39] $C_{\bar{\mu}} = \varepsilon_r \varepsilon_0 \kappa^2$.

Figure 1.6c depicts the quantum electrical circuit representation (red box) of the molecular films shown in Fig. 1.6a, which is expressed as the parallel sum of individual quantum resistors $(h/2e^2)$ and series capacitors $(C_{q,n})$, each coupled to an external capacitance (C_{ext}) associated with environmental effects and dielectrics. Note that in a vacuum, this is C_e, and in an electrolyte, it is equivalent to C_i, as discussed previously. Lastly, Fig. 1.6d illustrates the equivalent circuit of the ensemble depicted as series R_q and $C_{\bar{\mu}}$, which provides an electron transfer rate of $k = (R_q C_{\bar{\mu}})^{-1}$, following Eq. (1.15), where the length L corresponds to the thickness of the electron path through the molecular layer, as exemplified in Fig. 1.6.

Figure 1.6d shows, as previously demonstrated, that an ensemble of individual quantum point contacts exemplified by individual $h/2e^2$ and $C_{q,n}$ components, associated in series, possess an average response mesoscopically given by a series association of only R_q and $C_{\bar{\mu}}$. This is the response experimentally obtained by means of impedance spectroscopy measurements [21, 22, 25, 28, 41, 43]. The conductance $G = 1/R_q$ (Fig. 1.7b and d) and $C_{\bar{\mu}}$ (Fig. 1.7a) are experimentally accessible by impedance measurements [22] in alkane molecular films, where tunnelling and hopping mechanisms can be studied, as commonly reported by other methods [20, 21].

Figure 1.7 shows an experimental impedance spectroscopic measurement of molecular junctions consisting of alkane thiol monolayers. Figure 1.7a shows the Bode diagram representation of the real component of complex capacitance, while Fig. 1.7b shows the Bode diagram of AC conductance $[G(\omega) = Y'(\omega) = \omega C'']$, which is directly associated with the admittance of electrons to the molecular redox switch ensemble (see Eq. (3.2)).

Figure 1.7d shows the conductance obtained at the resonant frequency, 20 Hz (red vertical lines in Figs. 1.7a, b and c). The temperature dependence of $C_{\bar{\mu}}$ and G is shown in this figure to be consistent with the fact that an increase in capacitance concomitantly causes an increase in the conductance. Figure 1.7c demonstrates that this 20 Hz resonance frequency is temperature-independent and is defined as $k = G/C_{\bar{\mu}}$. Figure 1.7d also shows that G maximises when electrode energy is aligned with redox-active sites in the switches, corresponding to E_r.

It is noteworthy that the temperature independence of $k = G/C_{\bar{\mu}}$ (see Fig. 1.7c) is suggestive of the tunnelling mechanism and occurs even when G and $C_{\bar{\mu}}$ are temperature dependent within a given interval of frequencies. An extreme electron transfer rate depends on $C_{\bar{\mu}}$ and hence on the associated nature of the electronic density of states.

To finalise, we must emphasise that although $C_{\bar{\mu}}$ is a decisive factor in understanding the properties of mesoscopic systems, it is a component that has been ignored in molecular electronics, and only (albeit rarely) qualitatively inferred as highest occupied molecular orbital (HOMO) and lowest unoccupied molecular

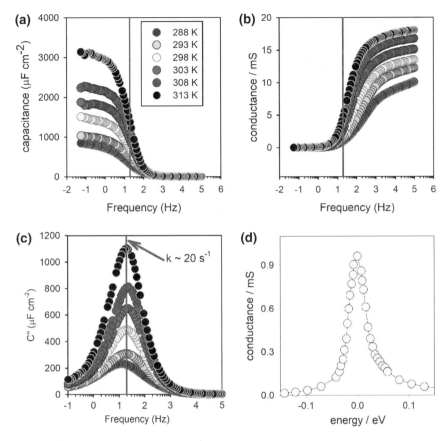

Fig. 1.7 **a** Bode representation of the real component of complex capacitance. **b** Bode diagram of AC conductance, $G(\omega) = \omega C''$, as a function of temperature. **c** Temperature dependence of C'' as a function of temperature. **d** G can be equivalently measured as a function of potential at a fixed frequency, and as predicted, maximises when electrode energy is aligned with a redox site. Finally, note that the resonant frequency occurs at 20 Hz. *Reprinted (adapted) with permission from Paulo R. Bueno, Tiago A. Benites and Jason J. Davis; The Mesoscopic Electrochemistry of Molecular Junctions, Scientific Reports. Copyright (2016) Nature*

orbital (LUMO) gaps [7, 94]. Nonetheless, $C_{\bar{\mu}}$ is a quantitative measurement of HOMO-LUMO states, as will be shown in Chap. 2 [41].

In the theoretical framework introduced here, the electron transfer rate depends explicitly on $C_{\bar{\mu}}$ within its quantum mechanics components, which differ from C_{μ}, in which the electrical field screening is different. It is because of the latter that we explicitly differentiate an electrochemical $d\bar{\mu} = -edV$ difference attained for a single-contact electrochemical measurement (see Fig. 1.1c) from the chemical potential (non-equilibrium) differences ($d\mu = -eV$) obtained in two-terminal configurations (Figs. 1.1a and b).

In summary, electrochemically gated or not, single- and two-terminal experimental set-ups differ. Such differences are due not only to issues involving chemical bonds to the electrodes (single or double contacts) but also, and mainly, to altered electrical field screening. In Chap. 3, we will discuss how Eq. (1.15) complies with frequency dependence, field effects and thermal broadening [22, 42] and also offer experimental examples.

Finally, in treating the molecular states electronically coupled with electrodes as point contacts, and further, in electrochemically measuring and interpreting the properties of the ensemble from a time-dependent standpoint, a correspondence is found between the Landauer [104] quantum of conductance and the electrochemical capacitance [29, 41] through the electron transfer rate [22, 105], which deeply alters the way in which electronics and electrochemistry are interpreted at the molecular level. Indeed, although there may be differences in the experimental set-up, there are no conceptual differences between electronics and electrochemistry at the molecular scale.

References

1. F.A. Buot, Mesoscopic physics and nanoelectronics—Nanoscience and nanotechnology. Phys. Rep.—Review Section of Phys. Lett. **234**(2–3), 73–174 (1993)
2. Y. Cui, Z.H. Zhong, D.L. Wang, W.U. Wang, C.M. Lieber, High performance silicon nanowire field effect transistors. Nano Lett. **3**(2), 149–152 (2003)
3. M.S. Gudiksen, L. J. Lauhon, J. Wang, D.C. Smith, C.M. Lieber, Growth of nanowire super-lattice structures for nanoscale photonics and electronics, Nature **415**(6872), 617–620 (2002)
4. W. Lu, C.M. Lieber, Nanoelectronics from the bottom up. Nat. Mater. **6**(11), 841–850 (2007)
5. Q.H. Wang, K. Kalantar-Zadeh, A. Kis, J.N. Coleman, M.S. Strano, Electronics and opto-electronics of two-dimensional transition metal dichalcogenides. Nat. Nanotechnol. **7**(11), 699–712 (2012)
6. A. Nitzan, M. A. Ratner, Electron transport in molecular wire junctions. Science **300**(5624), 1384–1389 (2003)
7. N.J. Tao, Electron transport in molecular junctions. Nat. Nanotechnol. **1**(3), 173–181 (2006)
8. J. Gabelli, G. Feve, J.M. Berroir, B. Placais, A. Cavanna, B. Etienne, Y. Jin, D.C. Glattli, Violation of Kirchhoff's laws for a coherent RC circuit. Science **313**(5786), 499–502 (2006)
9. R. Landauer. Electrical resistance of disordered one-dimensional lattices. Philos. Mag. **21**(172), 863–867 (1970)
10. R. Landauer. Future evolution of computer. Phys. Today **23**(7), 22 (1970)
11. Y. Gefen, Y. Imry, M.Y. Azbel, Quantum oscillations and the Aharonov-Bohm effect for parallel resistors. Phys. Rev. Lett. **52**(2), 129–132 (1984)
12. S. Ilani, L.A.K. Donev, M. Kindermann, P.L. McEuen, Measurement of the quantum capacitance of interacting electrons in carbon nanotubes. Nat. Phys. **2**, 687–691 (2006)
13. M. Büttiker, A. Thomas, A. Prêtre, Mesoscopic capacitors. Phys. Lett. A **180**(4–5), 364–369 (1993)
14. L.P. Kadanoff, G. Baym, *Quantum Statistical Mechanics: Green's Function Methods in Equilibrium and Nonequilibrium Problems* (W.A. Benjamin Inc, New York, 1962)
15. B.Q. Xu, X.Y. Xiao, X.M. Yang, L. Zang, N.J. Tao, Large gate modulation in the current of a room temperature single molecule transistor. J. Am. Chem. Soc. **127**(8), 2386–2387 (2005)
16. C.R. Arroyo, S. Tarkuc, R. Frisenda, J.S. Seldenthuis, C.H.M. Woerde, R. Eelkema, F.C. Grozema, H.S.J. van der Zant, Signatures of quantum interference effects on charge transport through a single benzene ring. Angew. Chem.-International Edition **52**(11), 3152–3155 (2013)

17. C. Li, A. Mishchenko, T. Wandlowski. Charge transport in single molecular junctions at the solid/liquid interface, in *Unimolecular and Supramolecular Electronics Ii: Chemistry and Physics Meet at Metal-Molecule Interfaces*, vol. 313, ed. by R.M. Metzger (2012), pp 121–188

18. L. Venkataraman, J.E. Klare, C. Nuckolls, M.S. Hybertsen, M.L. Steigerwald. Dependence of single-molecule junction conductance on molecular conformation. Nature **442**(7105), 904–907 (2006)

19. N.J. Tao, Probing potential-tuned resonant tunneling through redox molecules with scanning tunneling microscopy. Phys. Rev. Lett. **76**(21), 4066–4069 (1996)

20. W.C. Ribeiro, L.M. Goncalves, S. Liebana, M.I. Pividori, P.R. Bueno, Molecular conductance of double-stranded DNA evaluated by electrochemical capacitance spectroscopy. Nanoscale **8**(16), 8931–8938 (2016)

21. P.R. Bueno, J.J. Davis, Measuring quantum capacitance in energetically addressable molecular layers. Anal. Chem. **86**, 1337–1341 (2014)

22. P.R. Bueno, T.A. Benites, J.J. Davis. The mesoscopic electrochemistry of molecular junctions. Sci. Rep. **6**, 18400 (2016)

23. Y.Q. Xue, M.A. Ratner, Theoretical principles of single-molecule electronics: A chemical and mesoscopic view. Int. J. Quantum Chem. **102**(5), 911–924 (2005)

24. Λ.M. Kuznetsov, J. Ulstrup, *Electron Transfer in Chemistry and Biology. An Introduction to the Theory* (Wiley, Chichester, 1999)

25. P.R. Bueno, J.J. Davis, G. Mizzon, Capacitance spectroscopy: A versatile approach to resolving the redox density of states and kinetics in redox-active self-assembled monolayers. J. Phys. Chem. C. **116**(30), 8822–8829 (2012)

26. P.R. Bueno, C. Gabrielli. *Electrochemistry, Nanomaterials and Nanostructures* (Springer, New York, 2008)

27. P.R. Bueno, F. Fabregat-Santiago, J.J Davis, Elucidating capacitance and resistance terms in confined electroactive molecular layers. Anal. Chem. **85**(1), 411–417 (2013)

28. F.A. Gutierrez, F.C.B. Fernandes, G.A. Rivas, P.R. Bueno, Mesoscopic behaviour of multi-layered graphene: the meaning of supercapacitance revisited. Phys. Chem. Chem. Phys. **19**(9), 6792–6806 (2017)

29. D.A. Miranda, P.R. Bueno, Density functional theory and an experimentally-designed energy functional of electron density. Phys. Chem. Chem. Phys. **18**(37), 25984–25992 (2016)

30. A.J. Bard, L.R. Faulkner, *Electrochemical Methods: Fundamentals and Applications*, 2nd edn. (Wiley, New York, 2001)

31. N. Agrait, A.L. Yeyati, J.M. van Ruitenbeek, Quantum properties of atomic-sized conductors. Phys. Rep.—Review Section of Phys. Lett. **377**(2–3), 81–279 (2003)

32. M. Brandbyge, J.L. Mozos, P. Ordejon, J. Taylor, K. Stokbro. Density-functional method for nonequilibrium electron transport. Phys. Rev. B. **65**(16) (2002)

33. A.P. Alivisatos. Semiconductor clusters, nanocrystals, and quantum dots. Science **271**(5251), 933–937 (1996)

34. M. Bruchez, M. Moronne, P. Gin, S. Weiss, A.P. Alivisatos, Semiconductor nanocrystals as fluorescent biological labels. Science **281**(5385), 2013–2016 (1998)

35. W.C.W. Chan, S.M. Nie. Quantum dot bioconjugates for ultrasensitive nonisotopic detection. Science **281**(5385), 2016–2018 (1998)

36. X. Michalet, F.F. Pinaud, L.A. Bentolila, J.M. Tsay, S. Doose, J.J. Li, G. Sundaresan, A.M. Wu, S.S. Gambhir, S. Weiss, Quantum dots for live cells, in vivo imaging, and diagnostics. Science **307**(5709), 538–544 (2005)

37. C.W.J. Beenakker, H. Vanhouten, Quantum transport in semiconductor nanostructures. Solid State Phys. **44**, 1–228 (1991)

38. T. Mokari, E. Rothenberg, I. Popov, R. Costi, U. Banin, Selective growth of metal tips onto semiconductor quantum rods and tetrapods. Science **304**(5678), 1787–1790 (2004)

39. J. Lehr, J.R. Weeks, A. Santos, G.T. Feliciano, M.I.G. Nicholson, J.J. Davis, P.R. Bueno. Mapping the ionic fingerprints of molecular monolayers. Phys. Chem. Chem. Phys. (2017)

40. S. Luryi, Quantum capacitance devices. Appl. Phys. Lett. **52**, 501 (1988)

41. P.R. Bueno, G.T. Feliciano, J.J. Davis, Capacitance spectroscopy and density functional theory. Phys. Chem. Chem. Phys. **17**, 9375–9382 (2015)
42. A.L. Eckermann, D.J. Feld, J.A. Shaw, T.J. Meade, Electrochemistry of redox-active self-assembled monolayers. Coord. Chem. Rev. **254**(15–16), 1769–1802 (2010)
43. P.R. Bueno, D.A. Miranda, Conceptual density functional theory for electron transfer and transport in mesoscopic systems. Phys. Chem. Chem. Phys. **19**(8), 6184–6195 (2017)
44. J. Cecchetto, F.C.B. Fernandes, R. Lopes, P.R. Bueno, The capacitive sensing of NS1 Flavivirus biomarker. Biosens. Bioelectron. **87**, 949–956 (2017)
45. F.C.B. Fernandes, M.S. Goes, J.J. Davis, P.R. Bueno, Label free redox capacitive biosensing. Biosens. Bioelectron. **50**, 437–440 (2013)
46. J. Lehr, F.C.B. Fernandes, P.R. Bueno, J.J. Davis, Label-free capacitive diagnostics: exploiting local redox probe state occupancy. Anal. Chem. **86**(5), 2559–2564 (2014)
47. A. Santos, F.C. Carvalho, M.C. Roque-Barreira, P.R. Bueno, Impedance-derived electrochemical capacitance spectroscopy for the evaluation of lectin-glycoprotein binding affinity. Biosens. Bioelectron. **62**, 102–105 (2014)
48. A. Santos, J.P. Piccoli, N.A. Santos, E.M. Cilli, P.R. Bueno, Redox-tagged peptide for capacitive diagnostic assays. Biosens. Bioelectron. **68**, 281–287 (2015)
49. F.F. Hudari, G.G. Bessegato, F.C.B. Fernandes, M.V.B. Zanoni, P.R. Bueno, Reagentless detection of low-molecular-weight triamterene using self-doped TiO_2 Nanotubes. Anal. Chem. (2018)
50. E.P. Wigner, Lower limit for the energy derivative of the scattering phase shift. Phys. Rev. **98**(1), 145–147 (1955)
51. J.B. Goodenough, K.S. Park, The Li-ion rechargeable battery: a perspective. J. Am. Chem. Soc. **135**(4), 1167–1176 (2013)
52. Q. Wang, J.E. Moser, M. Gratzel, Electrochemical impedance spectroscopic analysis of dye-sensitized solar cells. J. Phys. Chem. B. **109**(31), 14945–14953 (2005)
53. A.S. Arico, P. Bruce, B. Scrosati, J.M. Tarascon, W. Van Schalkwijk. Nanostructured materials for advanced energy conversion and storage devices. Nature Materials **4**(5), 366–377 (2005)
54. B.C.H. Steele, A. Heinzel, Materials for fuel-cell technologies. Nature **414**(6861), 345–352 (2001)
55. P.R. Bueno; G.D. Schrott, P.S. Bonanni, S.N. Simison, J.P. Busalmen, Biochemical capacitance of geobacter sulfurreducens biofilms. Chemsuschem **8**(15), 2492–2495 (2015)
56. P. Poizot, S. Laruelle, S. Grugeon, L. Dupont, J.M. Tarascon, Nano-sized transition-metaloxides as negative-electrode materials for lithium-ion batteries. Nature **407**(6803), 496–499 (2000)
57. P. Simon, Y. Gogotsi, Materials for electrochemical capacitors. Nat. Mater. **7**(11), 845–854 (2008)
58. J.M. Tarascon, M. Armand. Issues and challenges facing rechargeable lithium batteries. Nature **414**(6861), 359–367 (2001)
59. S.S. Iqbal, M.W. Mayo, J.G. Bruno, B.V. Bronk, C.A. Batt, J.P. Chambers, A review of molecular recognition technologies for detection of biological threat agents. Biosens. Bioelectron. **15**(11–12), 549–578 (2000)
60. J.R. Lakowicz, Radiative decay engineering: biophysical and biomedical applications. Anal. Biochem. **298**(1), 1–24 (2001)
61. N. Nelson, C.F. Yocum, Structure and function of photosystems I and II. Ann. Rev. Plant Biol. **57**, 521–565 (2006)
62. S. Hammes-Schiffer, Theory of proton-coupled electron transfer in energy conversion processes. Acc. Chem. Res. **42**(12), 1881–1889 (2009)
63. Y. Qiao, S.J. Bao, C.M. Li, Electrocatalysis in microbial fuel cells-from electrode material to direct electrochemistry. Energy Environ. Sci. **3**(5), 544–553 (2010)
64. C.J. Brabec, N.S. Sariciftci, J.C. Hummelen, Plastic solar cells. Adv. Funct. Mater. **11**(1), 15–26 (2001)
65. D. Gust, T.A. Moore, A.L. Moore, Mimicking photosynthetic solar energy transduction. Acc. Chem. Res. **34**(1), 40–48 (2001)

66. F. Odobel, E. Blart, M. Lagree, M. Villieras, H. Boujtita, N. El Murr, S. Caramori, C.A. Bignozzi, Porphyrin dyes for TiO_2 sensitization. J. Mater. Chem. **13**(3), 502–510 (2003)

67. G.M. Crouch, D. Han, S.K. Fullerton-Shirey, D.B. Go, P.W. Bohn, Addressable direct-write nanoscale filament formation and dissolution by nanoparticle-mediated bipolar electrochemistry. Acs Nano **11**(5), 4976–4984 (2017)

68. N. Ebejer, A.G. Guell, S.C.S. Lai, K. McKelvey, M.E. Snowden, P.R. Unwin, Scanning electrochemical cell microscopy: a versatile technique for nanoscale electrochemistry and functional imaging. In *Annual Review of Analytical Chemistry*, vol 6, ed. by R.G. Cooks, J.E. Pemberton (2013), pp. 329–351

69. S. Lemay, H. White, Electrochemistry at the nanoscale: tackling old questions, posing new ones. Acc. Chem. Res. **49**(11), 2371–2371 (2016)

70. S.M. Oja, Y.S. Fan, C.M. Armstrong, P. Defnet, B. Zhang, Nanoscale electrochemistry revisited. Anal. Chem. **88**(1), 414–430 (2016)

71. S.M. Oja, Y.S. Fan, C.M. Armstrong, P. Defnet, B. Zhang, Nanoscale electrochemistry revisited, vol. 88 (2016, p. 414). Anal. Chem. **88**(12), 6628–6628 (2016)

72. S.M. Oja, M. Wood, B. Zhang, Nanoscale electrochemistry. Anal. Chem. **85**(2), 473–486 (2013)

73. H. Sugimura, K. Okiguchi, N. Nakagiri, M. Miyashita, Nanoscale patterning of an organosilane monolayer on the basis of tip-induced electrochemistry in atomic force microscopy. J. Vac. Sci. Technol. B **14**(6), 4140–4143 (1996)

74. S. Zaleski, A.J. Wilson, M. Mattei, X. Chen, G. Goubert, M.F. Cardinal, K.A. Willets, R.P. Van Duyne, Investigating nanoscale electrochemistry with surface- and tip-enhanced Raman spectroscopy. Acc. Chem. Res. **49**(9), 2023–2030 (1996)

75. T.V.P. Bliss, G.L. Collingridge, A synaptic model of memory—long-term potentiation in the hippocampus. Nature **361**(6407), 31–39 (1993)

76. N.C. Danbolt, Glutamate uptake. Prog. Neurobiol. **65**(1), 1–105 (2001)

77. B.K. Day, F. Pomerleau, J.J. Burmeister, P. Huettl, G.A. Gerhardt, Microelectrode array studies of basal and potassium-evoked release of L-glutamate in the anesthetized rat brain. J. Neurochem. **96**(6), 1626–1635 (2006)

78. E.N. Pothos, V. Davila, D. Sulzer, Presynaptic recording of quanta from midbrain dopamine neurons and modulation of the quantal size. J. Neurosci. **18**(11), 4106–4118 (1998)

79. H. Jeong, B. Tombor, R. Albert, Z.N. Oltvai, A.L. Barabasi, The large-scale organization of metabolic networks. Nature **407**(6804), 651–654 (2000)

80. A. Magasinski, P. Dixon, B. Hertzberg, A. Kvit, J. Ayala, G. Yushin, High-performance lithium-ion anodes using a hierarchical bottom-up approach. Nat. Mater. **9**(4), 353–358 (2010)

81. Q.F. Zhang, E. Uchaker, S.L. Candelaria, G.Z. Cao, Nanomaterials for energy conversion and storage. Chem. Soc. Rev. **42**(7), 3127–3171 (2013)

82. S.Y. Chung, J.T. Bloking, Y.M. Chiang, Electronically conductive phospho-olivines as lithium storage electrodes. Nat. Mater. **1**(2), 123–128 (2002)

83. B. Kang, G. Ceder, Battery materials for ultrafast charging and discharging. Nature **458**(7235), 190–193 (2009)

84. M. Prabu, S. Selvasekarapandian, A.R. Kulkarni, S. Karthikeyan, G. Hirankumar, C. Sanjeeviraja, Ionic transport properties of $LiCoPO_4$ cathode material. Solid State Sci. **13**(9), 1714–1718 (2011)

85. E. Laviron, AC polarograpy and faradaic impedance of strongly adsorbed electroactive species. 2. Theoretical-study of a quasi-reversible reaction in the case of Framkin isotherm. J. Electroanal. Chem. **105**(1), 25–34 (1979)

86. E. Laviron, AC polarography and faradaic impedance of strongly adsorbed electroactive species. 1. Theoretical and experimental-study of quasi-reversible reaction in the case of Langmuir isotherm. J. Electroanal. Chem. **97**(2), 135–149 (1979)

87. P. Gibot, M. Casas-Cabanas, L. Laffont, S. Levasseur, P. Carlach, S. Hamelet, J.M. Tarascon, C. Masquelier, Room-temperature single-phase Li insertion/extraction in nanoscale Li(x)FePO(4). Nat. Mater. **7**(9), 741–747 (2008)

88. S.C. Yin, H. Grondey, P. Strobel, M. Anne, L.F. Nazar, Electrochemical property: Structure relationships in monoclinic $Li_{3-y}V_2(PO_4)(3)$. J. Am. Chem. Soc. **125**(34), 10402–10411 (2003)

89. A.J. Bard, L.R. Faulkner, *Electrochemical Methods Fundamentals and Applications*, 2nd edn. (Wiley, New York, 2000)

90. S. Das Sarma, S. Adam, E.H. Hwang, E. Rossi, Electronic transport in two-dimensional graphene. Rev. Mod. Phys. **83**(2), 407–470 (2011)

91. E. McCann, V.I. Fal'ko, Landau-level degeneracy and quantum Hall effect in a graphite bilayer. Phys. Rev. Lett. **96**(8) (2006)

92. R. Horodecki, P. Horodecki, M. Horodecki, K. Horodecki, Quantum entanglement. Rev. Mod. Phys. **81**(2), 865–942 (2009)

93. A. Nitzan, A relationship between electron-transfer rates and molecular conduction. J. Phys. Chem. A **105**(12), 2677–2679 (2001)

94. A. Nitzan, Electron transmission through molecules and molecular interfaces. Ann. Rev. Phys. Chem. **52**, 681–750 (2001)

95. E. Wierzbinski, R. Venkatramani, K.L. Davis, S. Bezer, J. Kong, Y. Xing, E. Borguet, C. Achim, D.N. Beratan, D.H. Waldeck, The single-molecule conductance and electrochemical electron-transfer rate are related by a power law. Acs Nano **7**(6), 5391–5401 (2013)

96. E. Katz, I. Willner, Integrated nanoparticle-biomolecule hybrid systems: synthesis, properties, and applications. Angew. Chem. International Edition **43**(45), 6042–6108 (2004)

97. C. Mora, K. Le Hur, Universal resistances of the quantum resistance-capacitance circuit. Nat. Phys. **6**(9), 697–701 (2010)

98. F. Fogolari, A. Brigo, H. Molinari, The Poisson-Boltzmann equation for biomolecular electrostatics: a tool for structural biology. J. Mol. Recognit. **15**(6), 377–392 (2002)

99. M.S. Goes, H. Rahman, J. Ryall, J.J. Davis, P.R. Bueno, A dielectric model of self-assembled monolayer interfaces by capacitive spectroscopy. Langmuir **28**(25), 9689–9699 (2012)

100. A.Y. Grosberg, T.T. Nguyen, B.I. Shklovskii, Colloquium: the physics of charge inversion in chemical and biological systems. Rev. Mod. Phys. **74**(2), 329–345 (2002)

101. Y. Levin, Electrostatic correlations: from plasma to biology. Rep. Prog. Phys. **65**(11), 1577–1632 (2002)

102. P. Geerlings, S. Fias, Z. Boisdenghien, F. De Proft, Conceptual DFT: chemistry from the linear response function. Chem. Soc. Rev. **43**(14), 4989–5008 (2014)

103. P.R. Bueno, J.J. Davis, Elucidating redox level dispersion and local dielectric effects within electroactive molecular films. Anal. Chem. **86**(4), 1977–2004 (2014)

104. R. Landauer, Conductance from transmission—common-sense points. Phys. Scr. **T42**, 110–114 (1992)

105. R.A. Marcus, N. Sutin, Electron transfers in chemistry and biology. Biochim. Biophys. Acta **811**(3), 265–322 (1985)

Chapter 2
Electrochemistry and First Principles of Quantum Mechanics

In Chap. 1, we demonstrated that nanoscale electronics and electrochemistry both originate from the same physical background and are underpinned by chemical capacitance and its association with quantised relaxation resistance, thus defining the timescale of electronic processes that occur at the molecular scale.

In the presence of an electrolyte, which attenuates the electron dynamics, there is electrochemical capacitance, causing the electron dynamics to be governed by the electron transfer rate.

Thus, the fundamental energy framework at the nanoscale can be obtained by defining the chemical or electrochemical capacitance of the processes. Accordingly, in this chapter, we will demonstrate that the energy associated with charging chemical or electrochemical capacitive states can be derived from the first principles of quantum mechanics, using a density functional Hamiltonian.

2.1 Chemical Capacitance and Density Functional Theory

Chemical capacitance can be deduced from first principles of quantum mechanics, using a theoretical approach based on the density functional theory (DFT) and the associated Hamiltonian. This means that we can use accepted fundamentals and principles of quantum mechanics to deduce the chemical capacitance without making any phenomenological assumptions or empirical approximations that eventually lead to models that only serve to fit experimental data.

Although this requires a major effort involving tedious calculations and theoretical assumptions, it is important to demonstrate, at the fundamental level, the significance of chemical capacitance and its association with chemical reactivity indices, which is fully achieved in density functional theory [1–3]. Although this chapter is not indispensable reading for those not interested in examining the origin of these concepts, I personally find it useful to do so, since previous knowledge of quantum mechanics and DFT is necessary to accurately interpret data.

© The Author(s) 2018
P. R. Bueno, *Nanoscale Electrochemistry of Molecular Contacts*, SpringerBriefs in Applied Sciences and Technology, https://doi.org/10.1007/978-3-319-90487-0_2

It is very important to demonstrate that, whether an electrolyte is present or not (see previous chapter), chemical capacitance or electrochemical capacitance can be deduced from quantum mechanics, which provides proper meanings to parameters of the mesoscopic model that can be measured experimentally. Thus, it can be unequivocally established that chemical capacitance is a fundamental concept directly associated with the energy state of a mesoscopic system of interest, enabling it to be engineered based on quantum electronic concepts. In other words, if a mesoscopic entity, e.g. the capacitance pertaining to a nanoelectronic circuit or similarly to a nanoscale electrochemical circuit, could be measured experimentally, then that would be of great value to design molecular-based circuits. This is precisely the case of chemical (or electrochemical) capacitance.

As a first-principle quantum mechanics technique, we will use density functional theory (DFT), which is a powerful and interesting method for predicting the properties of matter at the nanoscale. In other words, DFT today is foremost among the tools of quantum mechanics employed in chemistry to predict reaction properties, using quantum computation to simulates chemical systems of interest. The reason for this is that, as will be demonstrated below, DFT gives direct theoretical meaning to well-known reactive indices that have been empirically used by chemists for decades, providing not only a useful computational chemistry tool but also a conceptual framework to rationalise chemical reactions.

In terms of computational effort, in the last few decades, DFT has been used successfully in place of wave function quantum mechanical methods. Based on DFT, instead of the system's wave function one determines its ground-state particle density, which, in principle, contains all the necessary information about a chemical or physical system, as demonstrated by the Hohenberg–Kohn theorems [4].

Furthermore, as stated previously, DFT now has a well-established potential for quantifying chemical systems and providing them with a theoretical and conceptual basis [2, 5–11]. This is the so-called conceptual DFT (also known as chemical or chemical reactivity DFT) [2, 5–11] that has been developed since the 1980s, inspired by Robert Parr and coworkers [2, 10].

Conceptual DFT constitutes a basis for the definition of a whole family of functions, such as *chemical hardness* and *chemical softness*. As will be demonstrated in detail, if chemical capacitance is deduced from DFT, then it can be used to define these functions. In other words, based on the system's chemical capacitance, we can determine its hardness or softness and begin to target electrochemical reactivity, for instance, using this chemical index. The usefulness of this approach still remains to be tested experimentally.

According to the terminology used in DFT, the fundamental variable associated with all observables is, of course, the electron density, i.e. $\rho = \sum_i \psi_i^2$, where ψ_i can be expressed as a linear combination of Kohn–Sham orbitals, rather than the wave function.

Over the years, chemists have devised numerous empirical atomic or molecular models to enable predictions or explanations of chemical behaviour and reactivity [6, 9]. One such empirical parameter is electronegativity, χ, defined by Mulliken as the average of ionisation potential I and electron affinity energies A, $\chi = 1/2(I + A)$ [2]. This empirically defined parameter has a theoretical meaning in DFT. The associated electron (or electronic) chemical potential, μ, is considered a constant and is defined in DFT as [2].

$$\mu = \left(\frac{\delta E}{\delta N}\right)_v = \left(\frac{\delta E}{\delta \rho}\right)_v = -\chi, \tag{2.1}$$

where E is the electronic energy of the system and N is the number of electron particles pertaining to a given N-electron system to be studied, which may be an atom, a molecule or a cluster, etc. Accordingly, μ is clearly identified as the negative of Mulliken electronegativity, which is highly meaningful in DFT.

In Eq. (2.1), thermal energy is ignored for the sake of convenience, and the theoretical meaning is that the nuclei of the atomic or molecular system are considered to possess a constant potential energy, i.e. the applied external potential (v) leads to an energy-related term that is constant. Lastly, note that ρ is N per unit of volume (the electron density itself), then the third part of Eq. (2.1) refers to the functional relationship between E and electron density, ρ. Note, also, that both v and ρ are functions of $\vec{r}$ (spatial coordinate vector of the electrons in the system), which is omitted here, for simplicity.

The theoretical definition of *chemical hardness*, η (in a conceptual DFT), has been shown [2] to be $\eta = \left(\delta^2 E/\delta\rho^2\right)_v = (\delta\mu/\delta N)_v = (\delta\mu/\delta\rho)_v$, corresponding to the derivative of the chemical potential with respect to the number of electronic particles in the chemical system of interest. The inverse of this function is

$$\sigma = \left(\frac{\delta\rho^2}{\delta^2 E}\right)_v = \left(\frac{\delta N}{\delta \mu}\right)_v = \left(\frac{\delta \rho}{\delta \mu}\right)_v, \tag{2.2}$$

which is termed the *chemical softness* of the system. The basic idea behind the use of DFT that leads to the definition of Eq. (2.2) and its inverse function (*chemical hardness*) is that, in every reaction of an atom, a molecule or a chemical system (when defined as an N-electron system) induced by an external potential perturbation, for instance, a proportional response to the perturbation, can be detected and measured.

We emphasise that the approach to be demonstrated below is actually an example of the use of these reactivity descriptors defined through conceptual DFT in terms of a capacitive nomenclature, which holds up quite well in dealing with mesoscopic systems and hence, with electronic (and electrochemical) circuits at the nanoscale. The latter are obviously very desirable in nanoelectronics and electrochemistry, as explained in Chap. 1. This approach will also be important in understanding the electrochemical reactions that take place in molecular switches, discussed later herein, and is also fundamental in applications comprising molecular junctions at the nanoscale used in molecular diagnostics. These applications will serve as examples of the concepts developed in Chap. 1 and further discussed in this chapter, while the purpose of Chap. 3 will be to demonstrate the universality of the concepts introduced in this book.

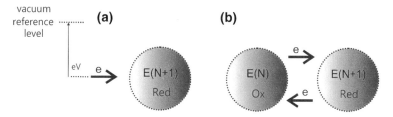

Fig. 2.1 a Introduction of an electron into an N-electron system (atomic or molecular) from vacuum. **b** An analogous representation of a self-exchange electron transfer between oxidised and reduced states of an N-electron system

Figure 2.1 schematically depicts an electron transfer or an energy storage process at the atomic scale, where (in Fig. 2.1a) an electron is introduced into an N-electron system (atomic or molecular) from a reference level of vacuum energy. If an electron is introduced or removed in this situation, the magnitude of the associated energy change is well-known as electron affinity $[A(N)]$ in the former case or as ionisation potential $[I(N)]$ in the latter. Figure 2.1b, analogously, represents an electron transfer process that occurs between oxidised and reduced states in the N-electron system. The latter situation is of great interest in electrochemistry because it represents a self-exchange electron transfer process. Figure 2.1b also illustrates a particularly important situation, largely ignored in current electron transfer rate models, which does not consider the influence of the electronic structure for the occurrence of a self-exchange.

In the presence of an electrode, as previously exemplified in Fig. 1.4c, the oxidation and reduction reactions can occur through the electronic states of the electrode. This situation, albeit not straightforward, can also be correlated with first-principle quantum mechanics.

2.2 Capacitance at the Atomic Scale

The capacitance that specifies the work, in terms of an externally applied electric potential V, which is required to bring a fixed amount of charge dq from the reference level of vacuum energy (see Fig. 2.1a) to a particular atomic scale environment containing a given number of N-electron particles, is defined as

$$\frac{1}{C} \equiv \frac{dV}{dq}, \tag{2.3}$$

where $dq = edN$, e is the unit of charge and N is the number of electrons, each of them stated at a particular spatial coordinate, $\vec{r}$, within the volume of the chemical system. This is important because it determines a particular electron density configuration and associated electronic structure in a specific volume that defines the properties of the system, according to DFT premises.

Equation (2.3) is normally introduced in classical mechanics electromagnetism, but it can be applied beyond this field, based on suitable premises. In contrast to classical premises, i.e. to the electronic energy associated with charging macroscopic electrodes (as depicted in Fig. 1.4a), an equivalent process at the atomic or molecular scale (Fig. 1.4b) involves a concomitant change in the potential of the electrons contained in the system. Thus, the chemical potential of the system as a whole is altered, and the potential variation in Eq. (2.1) can now be obtained as

$$edV = \mu(N + dN) - \mu(N) \equiv d\mu. \tag{2.4}$$

By combining Eqs. (2.3) and (2.4), we can define $C_\mu = edq/d\mu$ as the capacitance associated with a change in the electronic configuration of a given N-electron system with respect to the vacuum energy level (Fig. 2.1a). The charging energy associated with a single electron transfer process, which is obviously important in electrochemistry, particularly when associated with an atomic or a molecular entity, is thus provided assuming that $dN = 1$, so that and then

$$\frac{e^2}{C_\mu(N)} = \mu(N + 1) - \mu(N). \tag{2.5}$$

Note that Eq. (2.5) represents the work needed to move a single electron from the reference level of vacuum energy to an atomic or molecular system contained in the vacuum, which it does by considering the changes on electronic energy. In other words, this is not the electrostatic energy stored in a capacitor of typical geometry, C_e, quantifiable as $e^2/2C_e$. The latter is the case where the atomic or molecular levels are considered as conductive spheres, represented by a model known as Jellium, i.e. where the system is a uniform electron gas. On the other hand, Eq. (2.5) represents the situation in which additional contributions beyond the geometrical spatial separation of charges are computed. Those additional contributions are associated with the nature of the electronic configuration and interactions per se.

Accordingly, the relative contribution of C_μ to the total equivalent capacitance depends on the sensitivity of the chemical potential to the orbital occupancy and on the size and geometric configuration of the system. Now, considering the variation in the total energy, $E(N)$, associated with the addition or removal of a single electron in this particular situation, in which an atomic N-electron system is considered and the associated geometry must be taken into account, one has [12, 13].

$$\mu(N) = E(N) - E(N - 1) \tag{2.6}$$

from which it can be observed that

$$I(N) = E(N - 1) - E(N) \tag{2.7}$$

and

$$A(N) = E(N) - E(N + 1), \tag{2.8}$$

where $I(N)$ and $A(N)$ are the ionisation and association energies of the N-electron system. Given that the finite difference method [14] states that Eq. (2.4) represents the difference between the ionisation potential and electron affinity energies, we have [15, 16]

$$\frac{e^2}{C_\mu(N)} = E(N - 1) - 2E(N) + E(N + 1) = I(N) - A(N). \tag{2.9}$$

And now, upon dividing the above by two to establish an exact comparison with the electrostatic energy, $e^2/2C_e$, we find the exact definition of chemical hardness through a capacitive analysis [2], $\eta = e^2/2C_\mu(N) = [I(N) - A(N)]/2 = 1/2(\delta\mu/\delta\rho)_v$, from which it can be readily observed that the softness is

$$\sigma = \frac{2C_\mu(N)}{e^2} = 2\left(\frac{\delta\rho}{\delta\mu}\right)_v, \tag{2.10}$$

which is directly proportional to the density of states, i.e. $\mathcal{N} = (\delta\rho/\delta\mu)_v$ for the situation described in Figs. 1.4b and c.

In other words, the chemical capacitance, $C_\mu(N)$, can be defined as the capacitance associated with charging an N-electron chemical system, which is a function of the number of electronic particles and the electronic configuration [17]. Thus, $C_\mu(N)$ is directly associated with the total energy and its variations as a function of the variation in the number of electron particles in a given geometry of the N-electron chemical system.

In Chap. 1, it was demonstrated that an equivalent associated electrochemical energy [Eq. (1.2b)] can be defined by considering an external potential and a particular electric field screening. In the latter case, an associated capacitance can be defined, known as electrochemical capacitance. Notably, electrochemical capacitance is experimentally accessible by means of electrochemical spectroscopic methods (see Chap. 3 for further details).

Accordingly, from a fundamental standpoint following the equivalent reasoning of Eq. (2.9), electrochemical reactivity can be defined using electronegativity, ionisation and *chemical softness* (or its inverse, *chemical hardness*) terms, based on conceptual DFT.

In Chap. 3, a more detailed explanation is given of electrochemical reactions that occur in electroactive molecular films assembled on metallic electrodes, considering the thermal broadening and ensemble characteristics of these systems. At this point, however, based on the schematics in Fig. 2.2b, we can only briefly exemplify what the equivalent circuit of an electrochemical reaction would be, hypothetically speaking, at the zero approximation. Since the energy levels are not thermalised, they provide only an approximate pictorial representation of how the energy levels are coupled

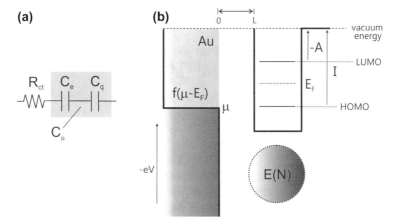

Fig. 2.2 a Equivalent circuit of the electrochemical activity of an electroactive quantum dot in contact with a metal electrode. **b** The quantum dot is represented as an N-electron system with an energy of $E(N)$

and how these levels can serve to predict the electrochemical reactivity of a quantum dot (or a molecule) in contact with an electrode.

In line with Eq. (1.15), Fig. 2.2a demonstrates the quantum channel resistance, $R_q = \sum h/2e^2 D$, and the electrochemical capacitance, $C_{\bar{\mu}}$. Then, the latter can be divided into electrostatic C_e and localised DOS (or quantum)-based contributions, C_q, as discussed in Chap. 1. Figure 2.2b shows an energy level model for a quantum dot (possessing multiple contacting channel modes) in contact to a metal plate electrode. The quantum dot is represented as an N-electron system with a total energy of $E(N)$.

In Fig. 2.2, note that the quantum dot "molecular" structure is approximated by a square well with molecular energy levels defined relative to the vacuum level. The frontier orbital energies are shown—correspondence with DFT is stated through the difference between HOMO and LUMO energy levels [$\varepsilon_{N+1} - \varepsilon_N$, according to Eq. (2.11), below]. The spatial separation between addressable quantum dot states and those of the electrode is given by L, as depicted in Fig. 2.2b. E_F is the Fermi energy of the quantum dot, $\bar{\mu}$ is the electrochemical potential of the electrons in the electrode. The capacitance of the junction and related electron coupling is then statistically governed by the Fermi–Dirac distribution function, $f(\bar{\mu} - E_F)$.

In the next sections, we continue to build upon these energy concepts within the context of DFT in order to definitely prove how $C_\mu(N)$ enables us not only to access chemical potential differences, defined through *chemical hardness* and *chemical softness* functions, but also to demonstrate how these functions connect and are resolved by solving the Hamiltonian. The solution of the Hamiltonian of an N-electron system, as defined in DFT, is important, which makes it essential to demonstrate the generality of chemical and electrochemical capacitance.

2.3   Chemical Capacitance and Kohn–Sham Eigenvalues

The importance of Eq. (2.9) stems from the fact that it can be correlated with Kohn–Sham eigenvalues as derived from DFT [12, 13], and this correlation is stated as

$$\frac{e^2}{C_\mu(N)} = E(N-1) - 2E(N) + E(N+1) = \varepsilon_{N+1} - \varepsilon_N + 2U_e, \qquad (2.11)$$

where ε_{N+1} and ε_N are LUMO and HOMO Kohn–Sham (KS) orbital energies and U_e is an approximately constant term associated with the external potential energy. Accordingly, in an isolated non-thermalised system, note that $e^2/C_q(N) = \varepsilon_{N+1} - \varepsilon_N$ is the quantum capacitance energy of the system, a term that is proportional to the *chemical hardness* and the *chemical softness* of a given atomic system, providing that the external potential is constant.

The implications of Eq. (2.11) for electrochemistry are significant; the quantum term of the electrochemical capacitance is implicitly given by the difference between ε_{N+1} (corresponding to the LUMO energy state) and ε_N (corresponding to the HOMO energy state) energies, plus a contribution from the external potential [18, 19]. Consequently, the capacitance of a chemical system such as the one illustrated in Fig. 2.2, comprising a contact between quantum dots or molecules and an electrode, is governed by the properties of the contact and of the electronic structure of the junction, which can be described by Kohn–Sham orbitals in DFT and thus be simulated based on numerical calculations.

Molecular examples of computational DFT calculations of the capacitance of systems and molecular films using computer-aided numerical simulations will be given later, in Sect. 2.6. Special attention will focus on the electronic wave function (calculated from DFT numerical methods), describing an electroactive molecular layer assembled on a gold electrode that forms a molecular junction, a singular and important building block in electronics and electrochemistry, as discussed in the introduction of this book. Equation (2.11) can lead to a continuum function that corresponds to the density of states (DOS) of the system when it is thermalised. This demonstrates that, when thermalised, $e^2/C_q(N) = \varepsilon_{N+1} - \varepsilon_N$ is a quantised energy directly proportional to the DOS.

In summary, this section demonstrated how chemical capacitance is significantly linked to DFT and also how it correlates to the DOS of a chemical system. The next sections will define a DFT Hamiltonian to be employed to solve the properties of a mesoscopic system. The aim is to demonstrate how the analytical solution of a DFT Hamiltonian correlates with the chemical capacitance of a mesoscopic system.

2.4 The Hamiltonian in Density Function Theory

Let us begin by recalling that, in principle, a quantum mechanical wave function contains all the information about a given system. By determining the wave function of the system, we can determine it's allowed energy states. However, it is unfortunately impossible to analytically solve the Schrödinger equation for an N-electron or N-body system. Evidently, albeit tricky, some approximations must be made to render the problem soluble. For this reason, we will adopt the simplest definition of DFT, which, for a chemist, is a method to obtain an approximate solution to the Schrödinger equation of a many-body system, which could be any atomic or many-electron systems such as nanoparticles, supramolecules, small proteins.

DFT, as a quantum mechanical method, finds the solution of the system by using electron density functionals. A functional is a function of a function. Thus, in DFT, the most basic functional is electron density, which is a function of space and time. Electron density is used in DFT as the fundamental property, unlike Hartree–Fock theory, which deals directly with the many-body wave function. By using electron density, DFT enables us to significantly speed up the calculation and simulation of chemical systems. In the DFT method, the total ground-state energy of a many-electron system is a functional of electron density. Hence, if the electron density functional is known, the total energy of the system can be determined a priori.

By focusing on electron density, it is possible to derive an effective one-electron-type Schrödinger equation. The total energy of the system can now be written in terms of a functional of the charge density.

Accordingly, in light of the DFT theory, the system Hamiltonian, in a fictitious system of non-interacting electrons, is written as a functional of spatial electron density, ρ, such as

$$E[\rho] = T_0[\rho] + \int_\Omega \rho(\vec{r})v(\vec{r})\mathrm{d}\vec{r} + \frac{1}{2}\iint_{\mho x \mho} \frac{e^2}{4\pi\varepsilon_r\varepsilon_0}\frac{\rho(\vec{r})\rho(\vec{r}')}{|\vec{r}-\vec{r}'|}\mathrm{d}\vec{r}\mathrm{d}\vec{r}' + E_{xc}[\rho] + E_{nn},$$

(2.12)

where the dependence of electron density on the spatial coordinates has been omitted in Eq. (2.12), for the sake of simplification. In the integrands, $\mho$ represents the spatial region where the N-electron system is located. T_0 is the kinetic energy, and the second and third terms are the interaction energy of the electrons with an external potential $[v(\vec{r})]$ and the electron–electron Coulomb energy, respectively. E_{xc} is the exchange-correlation energy and E_{nn} is the inter-nuclear Coulomb interaction energy.

The functional of Eq. (2.12) is minimised by the variational method under the constraint that the number of electrons N is constant (i.e. $N = \int \rho(\vec{r})v(\vec{r})\mathrm{d}\vec{r}$ is invariant) and is therefore carried out using the method of Lagrange multipliers:

$$\frac{\delta E[\rho]}{\delta \rho} - \varepsilon\left\{\frac{\delta}{\delta\rho}\left(\int_\Omega \rho(\vec{r})\mathrm{d}\vec{r} - N\right)\right\} = 0,$$

(2.13)

where ε is the Lagrange multiplier. In the particular situation, where the external Coulomb electric potential created by the nuclei is given by $v(\vec{r}) = \sum_a Z_a e^2/4\pi\varepsilon_r\varepsilon_0\left|\vec{r} - \vec{R}_a\right|$, Eq. (2.13) leads to the following equation

$$\int_\Omega \delta\rho \left\{ \frac{\delta T_0[\rho]}{\delta\rho} + \sum_a \frac{e^2 Z_a}{4\pi\varepsilon_r\varepsilon_0\left|\vec{r} - \vec{R}_a\right|} + \int \frac{e^2\rho(\vec{r}')}{4\pi\varepsilon_r\varepsilon_0\left|\vec{r} - \vec{r}'\right|} d\vec{r}' + \frac{\delta E_{xc}[\rho]}{\delta\rho} - \varepsilon \right\} = 0,$$

(2.14)

where the solution is found by assuming the sum of the terms in the brackets to be null. The only known expression for the kinetic energy is given in terms of the wave function, so the derivatives must be taken relative to the wave function rather than to the density. Taking the derivatives of this expression using the chain rule, the Kohn–Sham equation is [4, 20]

$$-\frac{\hbar^2}{2m}\nabla^2\psi_i + V_{\text{ef}}\psi_i = \varepsilon_i\psi_i,$$

(2.15)

where each ε_i for $i = 1,2,\dots N$ represents the value of each Lagrange multiplier, interpreted as the Kohn–Sham orbital energies, and V_{ef} is the Kohn–Sham effective potential given by

$$V_{ef} = \sum_a \frac{e^2 Z_a}{4\pi\varepsilon_r\varepsilon_0\left|\vec{r} - \vec{R}_a\right|} + \int_\Omega \frac{e^2\rho(\vec{r}')}{4\pi\varepsilon_r\varepsilon_0\left|\vec{r} - \vec{r}'\right|} d\vec{r}' + \frac{\delta E_{xc}[\rho]}{\delta\rho},$$

(2.16)

and the quantity $H_{KS} = -\left(\hbar^2/2m\right)\nabla^2 + V_{ef}$ is the Kohn–Sham Hamiltonian.

In summary, the Kohn–Sham equation is the one-particle Schrödinger equation, and the Kohn–Sham orbital energies are obtained as eigenvalues ε_i of the Kohn–Sham Hamiltonian.

2.5   Chemical Capacitance and the Functional of Electron Density

It would be useful to devise expressions that could deduce the capacitance of mesoscopic systems, using first-principle quantum mechanics. Naturally, this would require approximations and assumptions based on the definition of a mesoscopic system and the solution of the system's Hamiltonian. Let us start by defining a mesoscopic system, which could be a chemical system containing around 10^{12} atoms or molecules, for instance. This, then, assumes the existence of chemical systems containing a sufficient number of electrons to prevent us from solving the Hamiltonian accurately, based on computational calculations (the computational power currently available precludes this). However, these systems are not sufficiently large to allow for an approximation to classical mechanics, which is why quantum effects cannot be entirely neglected.

Having introduced the definition of mesoscopic systems, we will now solve the Hamiltonian of Eq. (2.12) considering the premise that a small perturbation of the system's electron density cannot significantly alter its energy state. This means that, upon applying a linear perturbation to the mesoscopic system, its electron density is perturbed only slightly in response to the removal or addition of a negligible number of electrons compared to the total number existing in the entire system. In so doing, one can see that the perturbation is linear and insignificant compared to the large number of electronic particles contained in the whole mesoscopic system, which could, for instance, be an assembly of molecules or a molecular film self-assembled on an electrode.

As a result, suppose we want to calculate the capacitance of a mesoscopic system that gets charged, that is, the total number of electrons changes from N to $N' = N + dN$. Naming $\rho'(\vec{r})$ the electron density of the charged system, the equation for the effective Kohn–Sham potential can be written as follows:

$$V'_{ef} = \frac{e^2}{4\pi\varepsilon_r\varepsilon_0} \sum_a \frac{Z_a}{|\vec{r} - \vec{R}_a|} + \int_\Omega \frac{e^2\rho'(\vec{r})}{4\pi\varepsilon_r\varepsilon_0|\vec{r} - \vec{r}'|} d\vec{r}' + \frac{\delta E_{xc}[\rho']}{\delta\rho'}. \qquad (2.17)$$

If one assumes that the geometry of the charged system does not change appreciably with respect to the neutral species, the difference between the Kohn–Sham potentials will be another constant, which can be written as:

$$V'_{ef} - V_{ef} = \frac{e^2}{4\pi\varepsilon_r\varepsilon_0} \left\{ \int \frac{\rho'(\vec{r})}{|\vec{r} - \vec{r}'|} d\vec{r}' - \int \frac{\rho(\vec{r})}{|\vec{r} - \vec{r}'|} d\vec{r}' \right\} + \left\{ \frac{\delta E_{xc}[\rho']}{\delta\rho'} - \frac{\delta E_{xc}[\rho]}{\delta\rho} \right\}. \qquad (2.18)$$

The electrostatic potential of a given charge distribution $\rho(\vec{r})$ can be expressed in terms of classical mechanics (see any textbook on electromagnetism) as $V = \frac{1}{4\pi\varepsilon_r\varepsilon_0} \int_\Omega \frac{\rho(\vec{r}')}{|\vec{r}-\vec{r}'|} d\vec{r}'$, which can be directly related to the capacitance, such that

$$\frac{e^2}{4\pi\varepsilon_r\varepsilon_0} \left\{ \int \frac{\rho'(\vec{r})}{|\vec{r} - \vec{r}'|} d\vec{r}' - \int \frac{\rho(\vec{r})}{|\vec{r} - \vec{r}'|} d\vec{r}' \right\} = \frac{(N' - N)e^2}{C}, \qquad (2.19)$$

which is interpreted as the "classical" (or geometric) capacitance.

Assuming that the variations in electron density are not extreme during the charging perturbation of the mesoscopic system, and that the exchange-correlation potential is independent of the gradient of electron density, it can be stated that

$$V'_{ef} - V_{ef} = \frac{(N' - N)e^2}{C}. \qquad (2.20)$$

Since the difference expressed in Eq. (2.20) is a constant (because it has been assumed that the external potential is constant), and given the structure of the Kohn–Sham Hamiltonian, the Kohn–Sham eigenvalues will be rigidly shifted by the same amount, and hence

$$\varepsilon_i' - \varepsilon_i = \frac{(N' - N)e^2}{C}. \tag{2.21}$$

In order to clarify what the classical and quantum contributions to capacitance would be, the total energy of the uncharged mesoscopic system can be rewritten, in terms of the Kohn–Sham eigenvalues, as

$$E[\rho] = \sum_i^N \varepsilon_i - \frac{1}{2}\frac{e^2}{4\pi\varepsilon_r\varepsilon_0} \iint_{\Omega\times\Omega} \frac{\rho(\vec{r})\rho(\vec{r}')}{|\vec{r}-\vec{r}'|}d\vec{r}d\vec{r}' + E_{xc}[\rho] - \int_\Omega \rho(\vec{r})\frac{\delta E_{xc}[\rho]}{\delta\rho(\vec{r})}d\vec{r} + E_{nn}. \tag{2.22}$$

Suppose that we want to calculate the capacitance of a mesoscopic system that gets charged, i.e. the total number of electrons change from N to N'. Keeping in mind that the electron density of the charged system was previously designated as $\rho'(\vec{r})$, the equation for the total energy of the charged system can now be expressed as

$$E[\rho'] = \sum_i^N \varepsilon_i' - \frac{1}{2}\frac{e^2}{4\pi\varepsilon_r\varepsilon_0} \iint_{\Omega\times\Omega} \frac{\rho'(\vec{r})\rho'(\vec{r}')}{|\vec{r}-\vec{r}'|}d\vec{r}d\vec{r}' + E_{xc}[\rho'] - \int_\Omega \rho'(\vec{r})\frac{\delta E_{xc}[\rho']}{\delta\rho'(\vec{r})}d\vec{r} + E_{nn}, \tag{2.23}$$

and hence, the total change in the energy of the system is

$$E[\rho'] - E[\rho] = \left(\sum_i^{N'} \varepsilon_i' - \sum_i^N \varepsilon_i\right)$$
$$- \frac{1}{2}\frac{e^2}{4\pi\varepsilon_r\varepsilon_0}\left(\iint_{\Omega\times\Omega} \frac{\rho'(\vec{r})\rho(\vec{r}')}{|\vec{r}-\vec{r}'|}d\vec{r}d\vec{r}' - \iint_{\Omega\times\Omega} \frac{\rho(\vec{r})\rho(\vec{r}')}{|\vec{r}-\vec{r}'|}d\vec{r}d\vec{r}'\right)$$
$$+ (E_{xc}[\rho'] - E_{xc}[\rho]) - \left(\int_\Omega \rho'(\vec{r})\frac{\delta E_{xc}[\rho']}{\delta\rho'(\vec{r})}d\vec{r} - \int_\Omega \rho(\vec{r})\frac{\delta E_{xc}[\rho]}{\delta\rho(\vec{r})}d\vec{r}\right). \tag{2.24}$$

In order to eliminate the explicit dependence of this expression on the coulomb and exchange-correlation terms, equations of the effective Kohn–Sham potential must be used. From Eqs. (2.18) and (2.20), one has

$$\frac{e^2}{4\pi\varepsilon_r\varepsilon_0}\left\{\int_\Omega \frac{\rho'(\vec{r}')}{|\vec{r}-\vec{r}'|}d\vec{r}' - \int_\Omega \frac{\rho(\vec{r}')}{|\vec{r}-\vec{r}'|}d\vec{r}'\right\} + \left\{\frac{\delta E_{xc}[\rho']}{\delta\rho'} - \frac{\delta E_{xc}[\rho]}{\delta\rho}\right\} = \frac{(N'-N)e^2}{C}. \tag{2.25}$$

If Eq. (2.25) is multiplied by $\frac{\rho(\vec{r})}{2}$ and integrated on $\vec{r}$, one has

$$\frac{e^2}{4\pi\varepsilon_r\varepsilon_0}\frac{1}{2}\iint\limits_{\Omega\times\Omega}\frac{\left[\rho'\left(\vec{r}'\right)-\rho\left(\vec{r}'\right)\right]\rho(\vec{r})}{\left|\vec{r}-\vec{r}'\right|}d\vec{r}d\vec{r}'$$

$$+\frac{1}{2}\int_{\Omega}\rho(\vec{r})\left(\frac{\delta E_{xc}[\rho']}{\delta\rho'}-\frac{\delta E_{xc}[\rho]}{\delta\rho}\right)d\vec{r}=\frac{1}{2}\int_{\Omega}\rho(\vec{r})\left(\frac{(N'-N)e^2}{C}\right)d\vec{r}=\frac{1}{2}\frac{N(N'-N)e^2}{C}.$$

$$(2.26)$$

Likewise, if Eq. (2.25) is multiplied by $\rho'(\vec{r})/2$ and integrated on $\vec{r}$, one has

$$\frac{e^2}{4\pi\varepsilon_r\varepsilon_0}\frac{1}{2}\iint\limits_{\Omega\times\Omega}\frac{\left[\rho'\left(\vec{r}'\right)-\rho\left(\vec{r}'\right)\right]\rho'(\vec{r})}{\left|\vec{r}-\vec{r}'\right|}d\vec{r}d\vec{r}'$$

$$+\frac{1}{2}\int_{\Omega}\rho'(\vec{r})\left(\frac{\delta E_{xc}[\rho']}{\delta\rho'}-\frac{\delta E_{xc}[\rho]}{\delta\rho}\right)d\vec{r}=\frac{1}{2}\int_{\Omega}\rho'(\vec{r})\left(\frac{(N'-N)e^2}{C}\right)d\vec{r}=\frac{1}{2}\frac{N'(N'-N)e^2}{C}.$$

$$(2.27)$$

Then, by adding up Eqs. (2.26) and (2.27), one has

$$\frac{e^2}{4\pi\varepsilon_r\varepsilon_0}\frac{1}{2}\iint\limits_{\Omega\times\Omega}\frac{\rho'\left(\vec{r}'\right)\rho'(\vec{r})}{\left|\vec{r}-\vec{r}'\right|}d\vec{r}d\vec{r}'$$

$$-\frac{e^2}{4\pi\varepsilon_r\varepsilon_0}\frac{1}{2}\iint\limits_{\Omega\times\Omega}\frac{\rho\left(\vec{r}'\right)\rho(\vec{r})}{\left|\vec{r}-\vec{r}'\right|}d\vec{r}d\vec{r}'=-\frac{1}{2}\int_{\Omega}\left[\rho(\vec{r})+\rho'\left(\vec{r}'\right)\right]\left(\frac{\delta E_{xc}[\rho']}{\delta\rho'}-\frac{\delta E_{xc}[\rho]}{\delta\rho}\right)d\vec{r}$$

$$+\frac{1}{2}\frac{(N'^2-N^2)e^2}{C}.$$

$$(2.28)$$

By substituting the result of Eq. (2.28) in Eq. (2.24), one obtains

$$E[\rho']-E[\rho]=\left(\sum_i^{N'}\varepsilon_i'-\sum_i^N\varepsilon_i\right)+\frac{1}{2}\int_{\Omega}\left[\rho(\vec{r})+\rho'\left(\vec{r}'\right)\right]\left(\frac{\delta E_{xc}[\rho']}{\delta\rho'}-\frac{\delta E_{xc}[\rho]}{\delta\rho}\right)d\vec{r}$$

$$-\frac{1}{2}\frac{(N'^2-N^2)e^2}{C}+\left(E_{xc}[\rho']-E_{xc}[\rho]\right)$$

$$-\left(\int_{\Omega}\rho'(\vec{r})\frac{\delta E_{xc}[\rho']}{\delta\rho'}d\vec{r}-\int_{\Omega}\rho(\vec{r})\frac{\delta E_{xc}[\rho]}{\delta\rho}d\vec{r}\right).$$

$$(2.29)$$

Equation (2.21) involving the Kohn–Sham eigenvalues can also be rearranged to obtain

$$\sum_i^{N'}\varepsilon_i'-\sum_i^N\varepsilon_i=\sum_i^{N'}\left[\varepsilon_i+\frac{(N'-N)e^2}{C}\right]-\sum_i^N\varepsilon_i=\sum_{N+1}^{N'}\varepsilon_i+\frac{N'(N'-N)e^2}{C}.$$

$$(2.30)$$

and thus, by substituting Eq. (2.30) in the first term of the right side of Eq. (2.29), we obtain

$$E[\rho'] - E[\rho] = \sum_{N+1}^{N'} \varepsilon_i + \frac{N'(N'-N)e^2}{C} + \frac{1}{2}\int_\Omega \left[\rho(\vec{r}) + \rho'(\vec{r})\right]\left(\frac{\delta E_{xc}[\rho']}{\delta\rho'} - \frac{\delta E_{xc}[\rho]}{\delta\rho}\right)d\vec{r}$$

$$- \frac{1}{2}\frac{(N'^2 - N^2)e^2}{C} + \left(E_{xc}[\rho'] - E_{xc}[\rho]\right)$$

$$- \left(\int_\Omega \rho'(\vec{r})\frac{\delta E_{xc}[\rho']}{\delta\rho'}d\vec{r} - \int_\Omega \rho(\vec{r})\frac{\delta E_{xc}[\rho]}{\delta\rho}d\vec{r}\right). \qquad (2.31)$$

The exchange-correlation terms can be eliminated by applying the following approximation:

$$\left(E_{xc}[\rho'] - E_{xc}[\rho]\right) = \frac{1}{2}\int_\Omega \left[\rho'(\vec{r}) - \rho(\vec{r})\right]\left\{\frac{\delta E_{xc}[\rho']}{\delta\rho'} + \frac{\delta E_{xc}[\rho]}{\delta\rho}\right\}d\vec{r} \qquad (2.32)$$

This holds whenever the variation in electron density is small compared to the total density of electronic states of the system, and usually involves situations that disturb the electron density of a variety of mesoscopic systems. Finally, upon substituting Eq. (2.32) by Eq. (2.31), one obtains the following expression:

$$E[N'] - E[N] = \sum_{N+1}^{N'} \varepsilon_i + \frac{(N'-N)^2 e^2}{2C} \qquad (2.33)$$

Applying this equation to the particular case, where $dN = \pm 1$, i.e. for $N' = N+1$ or $N' = N - 1$, Eq. (2.33) is simplified to assume the form of

$$E(N+1) - 2E(N) + E(N-1) = \varepsilon_{N+1} - \varepsilon_N + \frac{e^2}{C} \qquad (2.34)$$

Now the more general definition of the chemical capacitance can be recovered, which is equivalent to

$$\frac{e^2}{C} = \mu(N+1) - \mu(N) = E(N+1) - 2E(N) + E(N-1) \qquad (2.35)$$

This general definition not only explicitly illustrates the contribution of the quantum levels to variations in energy between charged and neutral states of a mesoscopic system, but also allows us to identify and separate the quantum from the total capacitive contribution (within electrostatic and quantum terms).

This result can be used not only to interpret experimentally measured chemical or electrochemical capacitance, which can be obtained via impedance spectroscopic methods [1], but also enables us to make direct comparisons with the computational calculations (see next section).

The right-hand side of Eq. (2.34) can also clarify the classic Coulomb contribution, which is clearly separate from the quantum capacitive contribution, at least in terms of its analytical and theoretical solution. These contributions cannot be separated in an experimental measurement of the chemical or electrochemical capacitance, but the theory helps us understand the workings of the different energy contributions. The quantised contribution is clearly associated with charging orbitals and accessible quantum states of the mesoscopic system, which can be observed by comparing Eqs. (2.34) and (2.35), from where the following expression can now be written:

$$\frac{e^2}{C_\mu(N)} = \frac{e^2}{C_e(N)} + \frac{e^2}{C_q(N)} \tag{2.36}$$

This expression defines the chemical capacitance, as introduced in Chap. 1, and also defines the electrochemical capacitance if the mesoscopic system is embedded in an electrolytic environment, which corresponds to Eq. (1.14). Lastly, Eq. (2.36) firmly demonstrates a correspondence with the quantum capacitance of the mesoscopic system, which, following only density functional methods, is now established as:

$$\varepsilon_{N+1} - \varepsilon_N = \frac{e^2}{C_q(N)}, \tag{2.37}$$

and is directly associated with the differences between LUMO and HOMO in a molecular mesoscopic system or molecular assembly. Consequently, the remaining term, e^2/C, is the $2U_e$ term of Eq. (2.11).

In summary, Eq. (2.34) is equivalent to Eq. (2.36), and hence, to $1/C_\mu = 1/C_e + 1/C_q$, *quod erat demonstrandum*, corresponding to a general situation that can be applied to any mesoscopic system. As this demonstration is based on first-principle quantum mechanics, it offers a general result. In other words, the approximation made for a mesoscopic physical scale follows suitable physical considerations that are valid, a priori, for any mesoscopic system, where variations (or perturbations) in electron density are negligible compared to the total electron density existing in the mesoscopic system.

We need to emphasise that in the presence of an electrolyte, the capacitance defined in Eq. (2.36) turns into that of Eq. (1.14), i.e. this is therefore the electrochemical capacitance. Depending on the environment, the electrochemical capacitance theoretically explains double-layer or pseudocapacitance, as generally referred to in the literature of electrochemistry. Double-layer or pseudocapacitance can be used as a probe for an underlying surface potential or local electrostatic change, and its sensitivity depends directly on the softness or hardness of a given molecular interface [21–23].

In the next section, the theory is tested within a DFT computational framework (using isolated and electrode coupled molecular systems) to observe the quantitative and qualitative impact of the electronic structure on the capacitance of mesoscopic systems. The computational results are also compared quantitatively with resolved experimental values obtained for electroactive molecular films.

2.6 Computational Density Functional Simulations of Molecular Films

In seeking to apply the theoretical fundamentals developed in the last section to the electrochemical capacitance obtained experimentally in electroactive molecular films assembled on metallic (gold) electrodes, the film's properties were simulated using a suitable atomistic model (see Fig. 2.3). The atomistic model comprises a gold cluster in which ferrocene-hexanethiol (a well characterised and commonly used electrochemical probe) is covalently attached to the metallic cluster. The quantum properties obtained in the system of the latter model were compared with those calculated for an analogous isolated ferrocenyl-hexanethiol system (the protonated form of the thiol). The latter has 40 atoms and is, of course, attached to a gold metal slab of 300 atoms.

After making DFT computational calculations, one can see (Fig. 2.3) that the positively charged gold cluster alkylthiolate-ferrocenium [$AuS\text{-}(CH_2)_6\text{-}Fc^+$] is the chemical entity (a mixture of charged and uncharged states exists in molecular films) that most contributes to the electrochemical capacitance associated with this junction. DFT geometries and their respectively calculated isodensities are depicted in Fig. 2.3 for both the LUMO (red) and HOMO (blue) states. The inset in this figure shows analogous depictions of the isolated molecules. Also, in this computationally simulated system, there is a notable energy state alignment within the surface-tethered configuration, demonstrating that there are superpositions of the electronic states of the molecular film on those contained in the gold electrode.

It is important to note that capacitance measurements can only be experimentally resolved for the surface-tethered film configuration, since only here can the dynamic

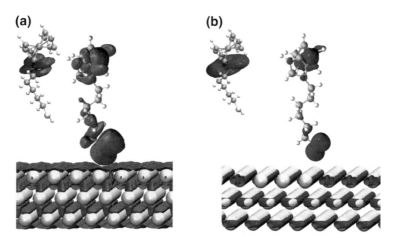

Fig. 2.3 Positively charged alkylthiolate-ferrocenium gold cluster [$AuS\text{-}(CH_2)_6\text{-}Fc^+$]. DFT geometries and respective calculated isodensities are shown for both the **a** LUMO (red) and **b** HOMO (blue) states

electrochemical reaction store energy and an associated capacitance be experimentally measured [24].

Nonetheless, in modelling or simulating the electronic structure of both isolated and tethered molecular configurations (Fig. 2.3), one can clearly observe how the electronic structure arising from the electrode–molecule coupling influences the capacitance by generating new energy states. These energy states exist only due to the molecule–electrode coupling and the contribution of this process to measurable charging associated with chemical capacitance.

In addition, by simulating the electronic structure of the cluster, it can be demonstrated that the experimentally resolved quantum capacitance aligns well with DFT generated predictions of contributing orbital states, which is in agreement with Eq. (2.11) and the Kohn–Sham orbital interpretation in DFT. The newly coupled electronic state established at the interface provides energy alignment and decreases the HOMO–LUMO separation, thereby contributing to increase the capacitance at this interface.

By means of Eq. (2.11), we initially analyse the variation of the total energy of the system with respect to charge variation/electron occupancy of the states of the molecular junction. To avoid unnecessary computational complexity, all the DFT calculations were considered in vacuum conditions and at zero absolute temperature, thus accessing the intrinsic properties (the quantised characteristic of the gold slab is observed in Fig. 2.4) of the molecular system without having to sample over a range of thermally accessible conformations.

Once an initial ground state is defined (by optimising geometry according to DFT code), the variation in the system's chemical potential with respect to electron exchange is accessible. The chemical potential of the system was evaluated by DFT across three states of charge, in line with Eq. (2.11), i.e. (i) the energy of the oxidised state, bearing N electrons, $E(N)$, here the Fe(III) state; (ii) the energy of the reduced state, containing $N + 1$ electrons, $E(N + 1)$, here the Fe(II) state; and (iii) the energy of $N - 1$ electron system, $E(N - 1)$. All the energies were calculated (see Table 2.1) within the same molecular geometry for both isolated and tethered molecular systems, with geometries (as shown in Fig. 2.3) assumed to be invariant throughout the electron exchange.

Here, it is worth noting again (Figs. 2.3 and 2.4) that it is the ferrocenium entity that most prominently contributes to C_q charging because it is here that HOMO and LUMO are mostly energetically aligned with underlying gold states. As shown, there is no evidence for iron and gold state energy alignment for the neutral species, confirming the lack of neutral redox state contribution to the HOMO and LUMO electronic structure that modulates the chemical capacitance.

Figure 2.4 shows the projected DOS for the [AuS-$(CH_2)_6$-Fc] system (Fig. 2.4a) compared to the [AuS-$(CH_2)_6$-Fc$^+$] (Fig. 2.4b) system, and their corresponding magnified views (Figs. 2.4c and d), indicating the contribution of each atom to each projected molecular orbital. The relevant peaks are those closest to the zeroed Fermi energy level, which serves as a reference level between occupied and unoccupied states. The occupied states correspond to the peaks on the right (positive), whereas the unoccupied ones are positioned to the left-hand side (negative). In Fig. 2.4, the

Table 2.1 DFT calculated energies for the N-electron system considered herein, for both free [(HS-$(CH_2)_6$-Fc)] and surface-confined [AuS-$(CH_2)_6$-Fc] configurations

	$E(N+1)$ (eV)	$E(N)$ (eV)	$E(N-1)$ (eV)	ε_{N+1} (eV)	ε_N (eV)	$C_\mu(N)$ (a F)
HS-$(CH_2)_6$-Fc	−3900.462	−3895.099	−3886.345	−6.712	−6.773	0.047184
AuS-$(CH_2)_6$-Fc	−273,235.642	−273,232.095	−273,227.833	−3.899	−3.911	0.223776

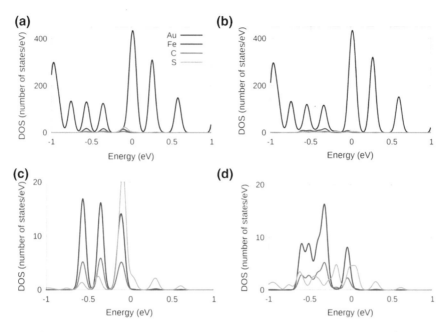

Fig. 2.4 Projected DOS for **a** the [AuS-(CH$_2$)$_6$-Fc] system and **b** the [AuS-(CH$_2$)$_6$-Fc$^+$] system, and the corresponding magnified views (**c**) and (**d**), showing the contribution of each atom to each projected molecular orbital

DOS was obtained from a convolution product between the Kohn–Sham eigenvalue spectrum and a Lorentzian broadening function of width 0.05 eV that are not strong enough to eliminate the observed quantised behaviour.

Figure 2.5a depicts the electronic structure of the isolated (positively charged) molecular system (geometry as shown in Fig. 2.3) together with the change associated with coupling to the metal slab (Fig. 2.5b). Most significantly, sufficient energy alignment was resolved here between the HOMO and LUMO states to allow the system to be described as having an effective DOS function. This effect is enhanced if the thermal broadening effects (as shown in Fig. 2.5b, continuous line) are integrated using Lorentz–Boltzmann statistics.

This increased chemical softness, in turn, increases the associated quantum capacitance [according to Eq. (2.37)]. It is important to note that the significant contribution to the energy alignment and coupling of gold orbitals of the "electrode" with iron orbitals of the attached molecular system containing ferrocene is enhanced by the 3d orbital of the iron itself and the carbon ring 2p states of the ferrocene (as evidenced in Fig. 2.3, when comparing the molecule–metal junction with the isolated molecule shown in the inset).

Figure 2.5 demonstrates in detail Kohn–Sham eigenvalue spectra (red lines), i.e. the energy states according to orbital analysis around the Fermi energy level for isolated ferrocene-hexanethiol (Fig. 2.5a) and for ferrocene-hexanethiol (Fig. 2.5b)

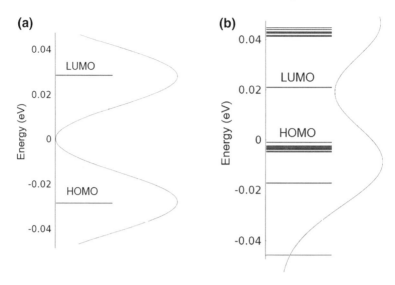

Fig. 2.5 DFT calculated energy states according to orbital analysis around the Fermi energy level for **a** isolated ferrocene-hexanethiol, and **b** for ferrocene-hexanethiol tethered to a metal slab

tethered to a metal slab. In Fig. 2.5b, there is evident perturbation of orbital states due to the presence of metallic states, with thermal broadening of the additional states introduced in the metallic-molecular junction between HOMO and LUMO of the system to room temperature, i.e. to 298 K. This can thus be described by a continuum DOS function, as shown in Fig. 2.5b. The decreased HOMO–LUMO energy gap and the increase in *chemical softness* of the system directly affect chemical capacitance, as discussed previously. The contributions of the accessible HOMO and LUMO quantum states to the capacitance follow Eq. (2.37).

Table 2.1 shows DFT calculated energies for the N-electron system considered herein, for both free [(HS-$(CH_2)_6$-Fc)] and surface-confined [AuS-$(CH_2)_6$-Fc] configurations. These energies were subsequently utilised, though Eq. (2.11), to calculate the electrochemical capacitance, C_μ. In Table 2.1, note that the Kohn–Sham orbital eigenvalues are expressed as ε_{N+1} and ε_N. Also, note that $E(N)$ corresponds to the total energy and ε_N to the energy of the specific eigenvalue. Finally, C_μ is calculated in (a) farads, where a is equivalent to 10^{-18}.

Using the energies obtained in Table 2.1, the quantum capacitance $e^2/C_q(N)$ contribution to the total capacitance was calculated by considering its correspondence with the differences between ε_{N+1} (LUMO eigenvalue) and ε_N (HOMO eigenvalue) of the molecular system following Eq. (2.37). Of course, the resolved quantum capacitance (Table 2.1) also has contributions that differ between the isolated and surface-tethered configurations, in a manner largely dependent on the occupancy of 3d-states associated with the ferrocene centre. The surface-tethered system is softer by virtue of the new orbital states generated (gold derived states) and has an increased quantum capacitance compared to the isolated system.

Gold metallic states that are energetically aligned with oxidised iron states in the molecule contribute to a reshaped molecular DOS and an enhanced charging efficiency (and thus enhanced C_μ). Significantly, the capacitive charge of the isolated molecule is both experimentally undetectable and energetically more demanding (see Table 2.1). In considering how sensitive the detectable capacitance is to orbital occupancy/electron density $[\delta\rho(N)/\delta\mu(N)]$ changes, it is instructive to note that this is expected to be greater the softer the systems, as is the particular case of the molecule–electrode system.[1]

Finally, the experimentally resolved quantum capacitive term (proportional to the DOS) obtained from electrochemical impedance experiments (Fig. 3.1) can be theoretically rationalised and quantified by DFT. Capacitive spectra shown in Fig. 3.1a and b correspond to measurements taken at the electrochemical half-wave potential where capacitance is maximised [14, 25, 26] (when the oxidised and reduced states responsible for capacitance are equally populated). A quantitative comparison with an experiment can be made by considering the particular situation in which Eq. (2.11) is applicable, resulting in $C_q = e^2 \mathcal{N}(E) = e^2 (\delta N/\delta E)_v$. Assuming that the majority of the states come from the quantised occupancy of the states, $C_q \sim C_\mu$, the experimentally resolved C_μ is highly dependent on the DOS situation theoretically predicted in Fig. 2.5b.

In making DFT calculations to include a more experimentally realistic depiction of molecular films, we used a cluster comprising 25 alkyl ferrocene molecules tethered to a metal slab (retaining geometries). Using these calculations we then found that $C_\mu = e^2 \mathcal{N}(E) \sim 230\mu\text{F cm}^{-2}$, which is in good quantitative agreement with the result found experimentally by capacitance-derived impedance spectroscopy for these films ($\sim$200 $\mu\text{F cm}^{-2}$—see Fig. 3.1a and b).

In this chapter, it was established that the energy associated with charging chemical capacitances (or electrochemical capacitance) can be derived from first principles of quantum mechanics, thus constituting a general and fundamental concept.

Furthermore, it was demonstrated that computationally simple DFT calculations outlined overlay satisfactorily with experimental observations (within capacitive-derived impedance spectroscopic experiments), since both are concerned with a consideration of parameters that are direct functions of the density of chargeable states, and again, are general within the DFT premises.

The density of the chargeable states was demonstrated to be that associated with Kohn–Sham frontier orbitals (the occupation of HOMO and LUMO frontier orbitals) with their respective eigenvalues. Since the electrochemical (or redox) characteristics of the surface-bound molecules are, of course, implicitly governed by the surface potential tuned alignment between these and electrode states, it follows that chemical (or electrochemical) capacitance is strongly influenced by the additional states introduced directly into the HOMO–LUMO gap during surface assembly.

---

[1]Note that softness was considered herein in terms of the electrochemical occupancy rather than polarisability of the electrochemical accessible centre or redox molecular switches, following the rationale given by Eq. (2.11).

An important DFT-supported observation is that charging efficiency correlates strongly with *chemical softness* of the orbital states. It was further shown that interfacial charging (of an electroactive film comprised of ferrocene molecules) is dominated by the oxidised ferrocene states because that is where energy alignment with metal states is optimal.

The theoretical framework presented here, based on conceptual DFT analysis is therefore general at any level of modern electronic structure theory (i.e. beyond DFT definitions of electronic structure). The electrochemical analysis it supports represents a general and valuable means of predicting experimental electrochemical observations of molecular films and lays the groundwork for an optimised approach in generating molecular films of either very high charging capacity or highly environmentally responsive charging.

References

1. P.R. Bueno, G.T. Feliciano, J.J. Davis, Capacitance spectroscopy and density functional theory. Phys. Chem. Chem. Phys. **17**, 9375–9382 (2015)
2. R.G. Parr, Y. Weitao, *Density-functional theory of atoms and molecules*. Oxford Science Publication (1994)
3. R.G. Parr, W.T. Yang, Density-functional theory of electronic-structure of molecules. Annu. Rev. Phys. Chem. **46**, 701–728 (1995)
4. P. Hohenberg, W. Kohn, Inhomogeneous electron gas. Phys. Rev. B **136**(3B), B864 (1964)
5. F. De Proft, P. Geerlings, Conceptual and computational DFT in the study of aromaticity. Chem. Rev. **101**(5), 1451–1464 (2001)
6. P. Senet, Kohn-Sham orbital formulation of the chemical electronic responses, including the hardness. J. Chem. Phys. **107**(7), 2516–2524 (1997)
7. H. Chermette, Chemical reactivity indexes in density functional theory. J. Comput. Chem. **20**(1), 129–154 (1999)
8. P.W. Ayers, J.S.M. Anderson, L.J. Bartolotti, Perturbative perspectives on the chemical reaction prediction problem. Int. J. Quantum Chem. **101**(5), 520–534 (2005)
9. B.E. Conway, *Electrochemical supercapacitors: scientific fundamentals and ttechnological applications*. (Springer, Berlin, 1999)
10. R.G. Parr, W.T. Yang, Density-function theory of electronic-structure of molecules. Annu. Rev. Phys. Chem. **46**, 701–728 (1995)
11. P. Senet, Nonlinear electronic responses, Fukui functions and hardnesses as functionals of the ground-state electronic density. J. Chem. Phys. **105**(15), 6471–6489 (1996)
12. G.J. Iafrate, K. Hess, J.B. Krieger, M. Macucci, Capacitive nature of atomic-sized structures. Phys. Rev. B **52**(15), 10737–10739 (1995)
13. J. Luo, Z.Q. Xue, W.M. Liu, J.L. Wu, Z.Q. Yang, Koopmans' theorem for large molecular systems within density functional theory. J. Phys. Chem. A **110**(43), 12005–12009 (2006)
14. P.R. Bueno, J.J. Davis, Measuring quantum capacitance in energetically addressable molecular layers. Anal. Chem. **86**(3), 1337–1341 (2014)
15. N.J. Tao, Probing potential-tuned resonant tunneling through redox molecules with scanning tunneling microscopy. Phys. Rev. Lett. **76**(21), 4066–4069 (1996)
16. W.C. Ribeiro, L.M. Goncalves, S. Liebana, M.I. Pividori, P.R. Bueno, Molecular conductance of double-stranded DNA evaluated by electrochemical capacitance spectroscopy. Nanoscale **8**(16), 8931–8938 (2016)
17. M. Büttiker, H. Thomas, A. Prêtre, Mesoscopic capacitors. Phys. Lett. A **180**, 364–369 (1993)

18. D.A. Miranda, P.R. Bueno, Density functional theory and an experimentally-designed energy functional of electron density. Phys. Chem. Chem. Phys. **18**(37), 25984–25992 (2016)
19. P.R. Bueno, D.A. Miranda, Conceptual density functional theory for electron transfer and transport in mesoscopic systems. Phys. Chem. Chem. Phys. **19**(8), 6184–6195 (2017)
20. W. Kohn, L.J. Sham, Self-consistent equations including exchange and correlation effects. Phys. Rev. **140**(4A), 1133 (1965)
21. F.C.B. Fernandes, M.S. Goes, J.J. Davis, P.R. Bueno, Label free redox capacitive biosensing. Biosens. Bioelectron. **50**, 437–440 (2013)
22. F.C.B. Fernandes, A. Santos, D.C. Martins, M.S. Goes, P.R. Bueno, Comparing label free electrochemical impedimetric and capacitive biosensing architectures. Biosens. Bioelectron. **57**, 96–102 (2014)
23. J. Lehr, G.C. Hobhouse, F.C.B. Fernandes, P.R. Bueno, J.J. Davis, Label-free capacitive diagnostics: exploiting local redox probe state occupancy (vol. 86, p 2559). Anal. Chem. **86**(7), 3682–3682 (2014)
24. P.R. Bueno, T.A. Benites, M.S. Goes, J.J. Davis, A facile measurement of heterogeneous electron transfer kinetics. Anal. Chem. **85**(22), 10920–10926 (2013)
25. P.R. Bueno, J.J. Davis, Elucidating redox level dispersion and local dielectric effects within electroactive molecular films. Anal. Chem. **86**(4), 1977–2004 (2014)
26. P.R. Bueno, G. Mizzon, J.J. Davis, Capacitance spectroscopy: a versatile approach to resolving the redox density of states and kinetics in redox-active self-assembled monolayers. J. Phys. Chem. B **116**(30), 8822–8829 (2012)

Chapter 3
Field Effect and Applications

Chapter 1 discussed the importance of establishing a unified standpoint for electronics and electrochemistry, and also showed that this can be done by suitably defining chemical or electrochemical capacitances. In Chap. 2, a detailed description was given about how chemical or electrochemical capacitances (if an electrolyte is considered in the analysis) can be derived using the first principles of quantum mechanics.

Thus, hopefully, readers are convinced about the importance of the preceding generalisations. Based on this expectation, in this chapter we confirm that classic electrochemistry, as an approximation of quantum mechanics, cannot depart from it by using classical electric circuit elements in its analysis of phenomena. Instead, it must use those that are defined in Chap. 1, adopting quantum principles and subsequent thermalisation of the circuit elements using ensembles and physical statistics.

This chapter also explains how the use of suitable circuit elements enables us to correctly interpret the properties of molecular electrochemical switches used as quantum sensor elements in biosensor applications, particularly in molecular diagnostics. Also demonstrated is how the field effect operates in these redox switches, and how they are therefore used as mesoscopic transistor elements of terminals. In addition, this chapter describes how circuit elements of electrochemical capacitance comply with supercapacitance phenomena in mesoscopic structures of graphene and titanium oxide. For the reasons previously stated, the compliance was obviously based on the analysis of the timescale of the process.

3.1 Faradaic and Non-Faradaic Processes

When discussing electronics and electrochemistry comparatively from a historical perspective, it is important to recall Michael Faraday's belief that electronics and electrochemistry are not different disciplines. Nonetheless, physicists appropriated Faraday's law of induction, which was important in building electrically powered

© The Author(s) 2018   51
P. R. Bueno, *Nanoscale Electrochemistry of Molecular Contacts*, SpringerBriefs
in Applied Sciences and Technology, https://doi.org/10.1007/978-3-319-90487-0_3

apparatuses as a separate branch. In contrast, the different fields of chemistry began to develop the discipline of electrochemistry, and here Faraday's constant [1] and Faraday's law of electrolysis [1] were fundamental to the progress achieved in this field. Thereafter, two different approaches to the phenomenon of capacitance, also introduced by Faraday (the unit of capacitance, Farad, was a tribute to him), were applied in electronics and electrochemistry.

Distinct interpretations of capacitance and capacitors, as elements of circuits in electronics and electrochemistry, have given rise to separate terminologies, i.e. electrostatic and electrochemical capacitors (double layer, for instance). In the field of electrochemistry, also in tribute to Faraday's law of electrolysis, the processes occurring in the presence or absence of charge transfer, respectively, are termed Faradaic and non-Faradaic [2–4].

In other words, as discussed in Chap. 1 and as will be reinforced with some experimental examples in this chapter, Faradaic and non-Faradaic processes are interfacial mechanisms that do or do not comply with Faraday's law of electrolysis. The differences are only a matter of how the electric field is associated with the charged particles that comprise the interface and how this field is screened at the junction by an electrolyte (this is discussed in detail in Sect. 3.6).

Thus, this chapter reinforces that there are no conceptual and physical differences between the charging processes that occur in electronics and electrochemistry, provided they are at the nanoscale. Electrochemical devices have the additional contribution of ions in the charging of their capacitive components, but this can be integrated in the theory, as demonstrated in Chap. 1. Accordingly, although the electrolyte is additional to electrochemical devices, it can be integrated with the theory that underpins mesoscopic electronic devices. We will begin by exploring junctions using frequency-dependent methods such as impedance and discussing what drives the screening of the electric field in nanoscale interfaces in contact with an electrolyte.

3.2 Experimental Frequency-Dependent Methodology

By following the perturbation theory, that is the linear dynamic response of a molecular system (like that depicted in Fig. 3.1), the time-dependent properties are accessed by the impedance or admittance of the system. This can be done experimentally using a potentiostat.

It follows that the frequency-dependent capacitance is measured as $C_{\bar{\mu}}(\omega) = dq(\omega)/dV(\omega)$, which (in its complex form) is dependent on the perturbation frequency (ω), such that [5, 6]

$$C_{\bar{\mu}}(\omega) \approx C_{\bar{\mu}}^0 \left(1 + j\omega R_q C_{\bar{\mu}}^0\right) + O(\omega^2) \approx \frac{C_{\bar{\mu}}^0}{1 + j\omega R_q C_{\bar{\mu}}^0} \tag{3.1}$$

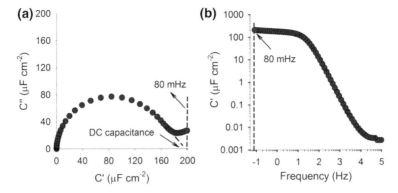

Fig. 3.1 **a** Capacitive Nyquist, and **b** Bode plots, showing that the electrochemical capacitance of a molecular junction, comprised of 11-ferrocenyl-undecanethiol self-assembled on a gold electrode, is about 200 μF cm^{-2}. *Reprinted (adapted) with permission from Paulo R. Bueno, Flavio C. B. Fernandes and Jason J. Davis; Quantum Capacitance as a Reagentless Molecular Sensing Element, Nanoscale. Copyright (2017) Royal Society of Chemistry*

where $\tau = 1/k = R_q C_{\bar{\mu}}^0$ (or $k = G/C_{\bar{\mu}}^0$) is the relaxation time of the charging process and $C_{\bar{\mu}}^0$ is the capacitance at $\omega \to 0$, so that $C_{\bar{\mu}}(\omega \to 0) = C_{\bar{\mu}}^0 = G/k$. The validation of Eq. (3.1) in terms of admittance, $G(\omega) = j\omega C_{\bar{\mu}}(\omega)$, can be confirmed experimentally by analysing its dissipative $Re[G(\omega)] = R_q$ and non-dissipative $Im[G(\omega)] = \omega C_{\bar{\mu}}$ terms, respectively.

Note also that $C_{\bar{\mu}}^0$, which is determined experimentally at a low frequency limit, as illustrated in Fig. 3.1, is not affected by the electron dynamics and this can be promptly ascertained, providing direct feedback about the electronic structure [5, 7] of the perturbed molecular ensemble, as demonstrated theoretically in Chap. 1. Furthermore, note that R_q is related to the real part of the AC conductance and therefore requires a dynamic theory. In analogy with Eq. (3.1), the conductance is expressed as

$$G(\omega) \approx -\omega C_{\bar{\mu}}^0 + \omega^2 \left(C_{\bar{\mu}}^0\right)^2 R_q + O(\omega^3), \tag{3.2}$$

for which, at the low frequency limit, the perturbation provides $G(\omega) \approx -\omega C_{\bar{\mu}}^0$.

Impedance-derived capacitance measurements (see Fig. 3.1) are taken at the Fermi level of a molecular quantum ensemble. This molecular quantum ensemble is comprised of 11-ferrocenyl-undecanethiol self-assembled on a gold electrode, forming a junction. The Fermi level of the junction coincides approximately with the electrochemical formal potential of the junction corresponding to the potential in the electrode, which coincides with the maximum of the electronic DOS curve depicted in Figs. 1.6b and 1.7d.

The direct graphical analysis in Fig. 3.1 is useful in determining the electrochemical capacitive behaviour, as discussed in the context of Eq. (3.1). This capacitance corresponds to the electrochemical capacitance (analysed theoretically in Chaps. 1 and 2) and can be measured, as illustrated in Fig. 3.1, as a function of potential scanned at a fixed predetermined frequency of 80 MHz, and the DOS curves can be constructed as exemplified in Figs. 1.6b and 3.8b.

Figure 3.1 clearly shows that once C_μ^0 is determined, then so is R_q, based on knowledge of the resonant frequency, $\omega_r = 1/R_q C_\mu^0$, which occurs at a frequency corresponding to the maximum of the capacitive semicircle-like response depicted in this figure. This is a direct measurement of the electron transfer rate, such as $k = \omega_r$, following the theory introduced in the context described by Eq. (1.15).

3.3   Field-Effect and Mesoscopic Electrochemical Transistors

It will now be demonstrated that a molecular ensemble of electrochemical capacitive point contacts operates analogously (albeit not equivalently, in that it follows quantum RC characteristics and the electric current is essentially dynamic) to a field-effect transistor format (short-circuited source and drain contacts) operating in an electrochemical environment.

Transistors [8–16] are devices used to amplify or switch electronic signals and electrical power. Field-effect transistors (FET) utilise an electric field to control the behaviour of the device and are typically made of three terminals, i.e. source (S), drain (D) and gate (G) (see Fig. 3.2a). The conductivity between the drain and source terminals is controlled by an electric field induced in the device, which is generated by the voltage differences between the channel and the gate terminals.

Chemically sensitive field-effect transistors [17–19] measure chemical alterations based on variations in the electric field of the chemically reactive environment in which they are embedded. For instance, when a target analyte concentration in a solution (in contact with the channel) is varied, the electric current passing through the transistor responds accordingly. A variation in the concentration of charged analyte ions in the solution induces a difference in chemical potential between the source and gate, which is measurable by FET as changes in the electric current. The design of this transistor is known as an electrolyte-gated FET [20, 21], because the ions do not penetrate into the channel, but instead, accumulate near its surface or near the surface of a dielectric layer (deposited on the channel), inducing accumulation of charge inside the channel, which is thus probed by electric field variations near and through the surface.

Organic electrochemical transistors [22–29] are a category of chemically sensitive FET in which the drain electric current is controlled by injecting ions from an electrolyte into the channel, altering the electronic charge density throughout its entire volume and resulting in very high transconductance. The disadvantage of organic electrochemical transistors is that, albeit sensitive, they are slow. The electronic properties in both electrolyte-gated FET and organic electrochemical transistors are controlled by ionic movement, and hence, by non-Faradaic characteristics of the

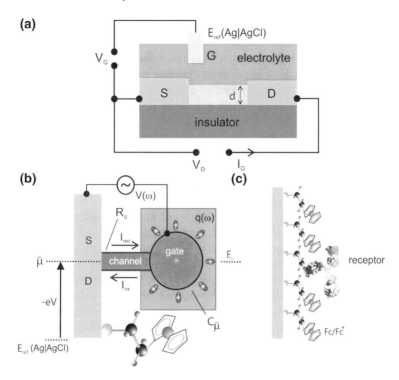

Fig. 3.2 a Typical architecture of an electrolyte-gated and chemically sensitive organic FET. **b** Pictorial representation of a mesoscopic electrochemical transistor molecularly designed using redox switches, i.e. ferrocene-terminated alkanethiols. **c** Illustration of redox-switched alkanethiol monolayers assembled on metallic electrodes within receptors, for molecular diagnostic applications

interface [18, 27]. The non-Faradaic charging characteristics of the interface are governed by double-layer capacitance and Debye field-effect screening, following the Debye length scale.

Figure 3.2a depicts a typical architecture of an electrolyte-gated and chemically sensitive organic FET, where E_{ref} refers to the electrochemical (silver/silver chloride) reference electrode, and d is the thickness of the organic (or dielectric) ion-permeable layer in contact with an electrolyte. Analogously, Fig. 3.2b contains a pictorial representation of a mesoscopic electrochemical transistor molecularly designed using redox switches, i.e. ferrocene-terminated alkanethiols (alternatively, a redox-active peptide chain can be used [30]; see further discussion in this chapter). Note that Figs. 3.2a, b are examples of electrochemical transistors, but Fig. 3.2b indicates that the mesoscopic electrochemical transistor is governed by quantum effects.

The goal here is to demonstrate how electrochemical capacitance, as a suitable molecular circuit element, can serve as a gate terminal in electrochemical transistors, representing a new class of electrochemical transistor. The nature of this category of transistor is intrinsically mesoscopic (Figs. 3.2b), combining classical and quantum mechanical effects within channel and gate components of the transistor, based on the concept of electrochemical capacitance.

As shown in Fig. 3.2b, the electric field in a mesoscopic electrochemical transistor is screened following redox rate dynamics. In other words, the dynamic equilibrium of the electrochemical current (oxidative and reductive), called exchange current [1], is resonant with electronic states contained in the electrode. The associated electrochemical capacitance, $C_{\bar{\mu}}(\omega) = dq(\omega)/dV(\omega)$, which is frequency dependent, can be ascertained by impedance-derived capacitance measurements [31].

In Fig. 3.2b, note that the source and drain electric current contributions depend on the steady state bias $dV = d\bar{\mu}/e$ imposed by a potentiostat. The mesoscopic FET structure reaches electrochemical equilibrium at $\bar{\mu} = E_r$, i.e. when $\bar{\mu}$, which is the electrochemical potential of the electrons in the electrode, is energetically aligned with E_r, which, in turn, is the formal potential of the redox switches. Accordingly, following the analysis given in Chap. 1, the Fermi level is constant at the junction, meaning that despite the presence of a dynamic resonant current, there is no net electron flux (see Fig. 3.2b).

Following the energy schemes shown in Fig. 3.2b, the difference between the electrode and gate potentials is defined as $dV_G = (V - V_r) = -d\bar{\mu}/e$. The gate is constituted of any modified layer (mesoscopic in character) in which electrochemical or quantum states are resolvable by an AC perturbation, such that the potential in the bridge (or the channel of the FET), V_c, is dependent on the electron density in the channel, N_c, i.e.

$$V_c = V_G - eN_c/C_i, \tag{3.3}$$

where $q_c = eN_c$ is the charge in the bridge and eN_c/C_i is the potential of the electrons in the bridge (screened by the electrolyte). The capacitance in the gate can thus be written as

$$\frac{dq_c}{dV_G} = \frac{dq_c}{dV_c}\frac{dV_c}{dV_G}, \tag{3.4}$$

Now, by applying the derivative of Eq. (3.3) with respect to V_G, we find that $dV_c/dV_G = 1 - [(1/C_e)(dq_c/dV_G)]$, allowing Eq. (3.4) to be rewritten as

$$\frac{dq_c}{dV_G} = \frac{dq_c}{dV_c}\left(1 - \frac{1}{C_i}\frac{dq_c}{dV_G}\right). \tag{3.5}$$

By assuming that dq_c/dV_G, the total capacitance, is the equivalent capacitance (of two series; C_i and C_q) of the system, that is, $C_{\bar{\mu}}$, and also by noting that dq_c/dV_c is the quantum capacitance, C_q, we obtain

$$\frac{1}{C_{\bar{\mu}}} = \frac{dV_G}{dq_c} = \left(\frac{C_q + C_i}{C_q C_i}\right) = \frac{1}{C_i} + \frac{1}{C_q}. \tag{3.6}$$

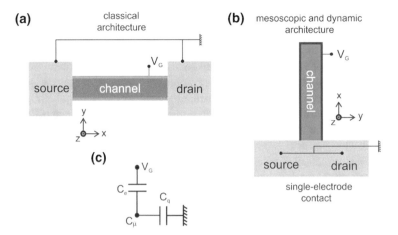

Fig. 3.3 **a** FET structure functioning in equilibrium DC conditions. **b** This architecture reflects the mesoscopic situation in which an external perturbation acts upon the system. **c** FET-associated equivalent circuit that governs behaviour when only gate voltage is varied in equilibrium conditions in configurations (**a** or **b**)

Finally, Eq. (3.6) demonstrates that a two-dimensional electroactive molecular ensemble behaves as a particular type of field-effect transistor. However, unlike traditional field-effect transistors (Fig. 3.3a), the properties of the gate can only be accessed dynamically because of the single-contact configuration, as shown in Fig. 3.3b. Equation (3.6) is equivalent to Eq. (1.14) and thus demonstrates that electrochemical capacitance can be used as circuit elements of molecular structures to design molecular transistors. This very important conclusion, which is in agreement with the theory introduced in Chaps. 1 and 2, affects the development of molecular junctions for electronic or electrochemical applications.

Dynamic measurements can be taken using time-dependent techniques such as impedance spectroscopy. The closest this analysis brings one to traditional transistors is if one considers an analogy where traditional field-effect transistors could operate under equilibrium conditions and an imposed AC potential is applied to investigate the equilibrium properties of the channel and gate. However, even this situation is not totally equivalent because of the way contacts are established in traditional transistors.

For instance, Fig. 3.3a illustrates a FET with a nanoscale channel structure (a molecular wire, for instance) functioning under equilibrium DC conditions, i.e. where the potential of the left- and right-hand electrodes are equal. Figure 3.3b shows a single-contact architecture that reflects the mesoscopic state where an external perturbation acting on the system, e.g. electric time-dependent fluxes from AC perturbation methods, is able to access the properties of the channel and the gate terminals. These FET architectures are similar inasmuch as they are both based on field effects, but they have essential differences that depend on the scale in which the FET device is operating, i.e. microscopic versus mesoscopic.

The electrochemical information contained in a mesoscopic and dynamic architecture can be obtained only from time-dependent electrical measurements. Figure 3.3c illustrates the FET-associated equivalent circuit that governs the behaviour when only gate voltage (V_G) is varied in equilibrium conditions, i.e. when the source and drain are poised at the same potential. In such a configuration, the equivalent capacitance is a chemical or electrochemical capacitance, measured as $C_\mu = C_{ext}C_q/(C_{ext} + C_q)$, where C_{ext} is the external capacitance, in agreement with Fig. 1.6a and Eq. (1.14). The only difference between $C_\mu = C_{ext}C_q/(C_{ext} + C_q)$ and Eq. (1.14) is how C_{ext} is defined. The chemical capacitance will be translated into electrochemical capacitance in the presence of an electrolyte, thus making C_{ext} will be a kind of ionic capacitance, such as a double layer. In the case of solid-state electronic circuits, C_{ext} is replaced by geometric or electrostatic capacitance.

In other words, the problem can be generalised and the difference between the potential in the channel (V_c) and in the gate (V_G) is dependent on $\mathcal{N}$ as $V_c - V_G = -(e\mathcal{N}/C_{ext})$. The gate capacitance is then $dq/dV_G = (dq/dV_c)(dV_c/dV_G)$. By noting that dq/dV_G is C_μ^0 and dq/dV_c is the quantum capacitance (C_q), dq/dV_G becomes rearranged as:

$$\frac{q^2}{C_\mu^0} = q^2\left(\frac{dV_G}{dq}\right) = q^2\left(\frac{C_q + C_{ext}}{C_q C_{ext}}\right) = q^2\left(\frac{1}{C_{ext}} + \frac{1}{C_q}\right) \tag{3.7}$$

which is an expression explicitly equivalent to Eq. (1.14), where q is the total electric charge involved in the process of charging the equivalent capacitance.

Thus, Eq. (3.7) implicitly demonstrates that, in experimentally accessing C_μ^0, the thermodynamic properties of the gates are traceable, causing molecular sensitive changes in the electric potential of the gates to become quantifiable. Consequently, the associated energy scale [5] of the gate is $E = q^2/2C_\mu^0$. Variations in this energy due to the occupancy of the states (in the gate or in the channel) associated with C_μ^0 can be reported through derivatives of E with respect to the (total) charge, thus resulting in $dE/dq = q/C_\mu^0$, which is an electric potential difference. For a single-electron transfer reaction at a constant temperature and pressure, $dE = dG = -edV_G = \Delta\bar{\mu}$, which is associated with the electrochemical potential difference between the gate and electrode [32].

Alternatively, in agreement with Eq. (2.35), it can be noted that $1/C_\mu^0$ is equivalent to dq/dV_G, where $qdV = \mu(N + dN) - \mu(dN) = d\bar{\mu}$ is the electrochemical potential difference due to the displacement of a number of electrons N. Assuming the exchange of a single electron $dN = 1$ and considering only the transfer of an elementary charge, this becomes $-edV_G = d\bar{\mu}$, which, when combined with $1/C_\mu^0 = dq/dV_G$, gives $e^2/C_\mu^0 = d\bar{\mu}$.

$d\bar{\mu} = e^2/C_\mu^0$ corresponds to the amount of energy spent in the transfer of a single electron from the electrode to a single quantum state in the gate, comprised of an ensemble of individual electrochemical switches.

At a finite temperature and under a fixed external potential (using same electrode potential and solvent environment), $d\bar{\mu}$ quantifies the differences between the HOMO and LUMO of the molecular ensemble [according to Eq. (2.11)], as depicted in Fig. 1.6b and 2.5a. Furthermore, based on the former analysis, it can be concluded that the ensemble of molecular contacts is a distribution of individual non-interacting electronic systems, equivalent to a two-dimensional electron gas [33].

In summary, in this section we demonstrated that electrochemical capacitance, in addition to being obtained experimentally using impedance spectroscopic methods, can serve as a gate and therefore designed as a quantum or molecular circuit element assembled directly on an electrode to operate as an element of a particular type of transistor, i.e. in mesoscopic molecular transistors. The next section demonstrates how this settles with thermal broadening, i.e. how these transistors realistically respond at room temperature.

3.4 Thermal Broadening

Electrochemical reactions realistically operate at room temperature, so the zero-temperature approximation [of Eqs. (1.14), (3.6) or (3.7)] is impracticable. Since $C_\mu^0 \propto [d\mathcal{N}/d\bar{\mu}]$, the easiest way to assess the thermal broadening of an ensemble of parallel quantum capacitors (as shown in Fig. 1.6b) is to resort to the grand canonical ensemble.

Assuming that the electrons are non-interacting, the states available for occupation in the mesoscopic capacitive point contacts (that form the molecular ensemble) are governed by a derivative of the average number of capacitive contacts with respect to the electrochemical potential of electrons, $[d\langle\mathcal{N}\rangle/d\bar{\mu}]$. This potential, which can be determined following the variance in capacitor numbers of the thermal broadening (at a constant temperature and microscopic volume, $\eth$) such that [34]

$$\left(\frac{d\langle\mathcal{N}\rangle}{d\bar{\mu}}\right)_{T,\eth} = \frac{1}{k_B T}\langle\mathcal{N}\rangle(1 - \langle\mathcal{N}\rangle) \tag{3.8}$$

where $\langle\mathcal{N}\rangle = (1 + \exp[-eV/k_B T])^{-1}$ is the average number of quantum capacitors and where $-eV = E_r - \bar{\mu}$. Keep in mind that E_r is the Fermi level associated with the ensemble of quantum capacitive states individually in contact with the electrode, and at electrochemical equilibrium $\bar{\mu} = E_r$; hence, thus $\langle\mathcal{N}\rangle = 1/2$. Considering a number density of states (or a molecular coverage of the reservoir), Γ, which defines the number of quantum capacitors covering the surface of the electrode, C_μ^0 is then finally obtained as

$$C_\mu^0 = \left[\frac{e^2 \Gamma}{k_B T}\right]\langle\mathcal{N}\rangle(1 - \langle\mathcal{N}\rangle), \tag{3.9}$$

since $e/k_B = F/R$, where R is the ideal gas constant and F is the Faraday constant. A relationship between this mesoscopic model and classical electrochemistry is once again established so that, as expected, classical electrochemistry is contained in the quantum equivalent. Recall Chap. 1, which contains complementary discussions.

For instance, consider the quantum capacitive ensemble comprised of a monolayer encompassing individual molecular electroactive entities (Fig. 1.6a), i.e. each molecule as an individual quantum point contact. In Chap. 1, it was explained that a collection of individual molecules forms an ensemble on the surface of an electrode, and that its capacitive macroscopic response is experimentally accessible as $C_{\bar{\mu}}^0$ by impedance or potential sweep methods.

Given that, in the condition of electrochemical equilibrium, it is known that $\langle \mathcal{N} \rangle = 1/2$, it can be stated that in a molecularly modified electrode, the product $\langle \mathcal{N} \rangle (1 - \langle \mathcal{N} \rangle)$, which is part of Eq. (3.9), equates to 1/4. Accordingly, the equivalent capacitance of the molecular ensemble, at the Fermi level of the junction (i.e. the potential of equilibrium of the junction), is given by $C_{\bar{\mu}}^0 = e^2 \Gamma / 4RT$, which is in line with the traditional electrochemistry analysis, as previously discussed in Chap. 1.

Thus, Eq. (3.9) conforms to the classical electrochemical analysis (see Fig. 1.3 in Chap. 1). Briefly restated, it has been demonstrated (in Chap. 1) that in classical electrochemical transient-based methods, the exchange electrochemical current density is determined based on Laviron's formalism [35, 36] such that $j = \left(e^2 \Gamma / 4RT \right) s$, where $s = dV/dt$ is the scan rate in the case of transient potential scan perturbing methods [86–87], and the electrochemical capacitance is therefore the proportional term between current density and scan rate.

If a distribution of electronic states on the molecular side of the junction, $g(\bar{\mu})$, is considered, then $\langle \mathcal{N} \rangle = g(\bar{\mu})(1 + \exp[-eV/k_B T])^{-1}$ and the electrochemical capacitance is obtained by integrating the overall contributions of available states, such that

$$C_{\bar{\mu}}^0(\bar{\mu}) = q^2 \int_{-\infty}^{+\infty} g(\bar{\mu}) \left(\frac{d\langle \mathcal{N} \rangle}{d\bar{\mu}} \right) d\bar{\mu}$$

$$= \frac{q^2}{k_B T} \int_{-\infty}^{+\infty} g(\bar{\mu}) \langle \mathcal{N} \rangle (1 - \langle \mathcal{N} \rangle) d\bar{\mu}, \tag{3.10}$$

which, for a zero-temperature approximation, takes on the form of $C_{\bar{\mu}}^0(\bar{\mu}) = q^2 g(\bar{\mu})$. Finally, it should be noted that, whichever is taken into account, Eq. (3.9) or Eq. (3.10), the conductance of the system is directly associated with quantised amounts of $2e^2/h$ and, overall, the total conductance is directly proportional to $C_{\bar{\mu}}^0(\bar{\mu})$, which is $G(\bar{\mu}) = kC_{\bar{\mu}}^0(\bar{\mu})$, and is now a function of $\bar{\mu}$ as driven by the potential of the electrode.

Countless applications and electrochemical devices can be developed following the reasoning previously introduced through Eqs. (3.9) or (3.10), simply because the capacitance and the associated thermodynamic statistics behaviour related to it can be projected as gates of a FET.

In summary, Eqs. (3.9) and (3.10) theoretically govern the nanoscale electrochemistry of a molecularly modified electrode. In the following sections, more applications will be described of this previously explained mesoscopic electrochemical device that typically governs the properties of molecular junctions. The theory will be applied to different molecular devices, obviously aiming to generalise the applicability of the described concepts and to reinterpret classical electrochemistry using a suitable quantum approach.

3.5 Envisaging Debye and Thomas–Fermi Screenings Experiment

Earlier in this book, we explained that both chemical and electrochemical capacitances are universal capacitances that consist, respectively, of electrostatic and ionic (resembling double layer) capacitances. The physical interpretation considers that electrostatic or double-layer comparable capacitances are associated in series with quantum capacitance.

The differences between these capacitances are thus dependent solely on the absence (Fig. 1.2a) or presence (Fig. 1.2b) of an ionic solvent structure, such as an electrolyte. Therefore, in the presence of an electrolyte, ions play an important role in the capacitive charging phenomena of an electrode/electrolyte interface. C_{μ}^0 contains a C_i component that outweighs C_q in the *absence* of redox-active components in an electroactive layer (or electronic accessible states). In other words, C_i is ruled by non-Faradaic effects [1] in which ion-associated Debye screening prevails. On the other hand, if there are redox-accessible states, then C_{μ}^0 includes the influence of the electron accessible states, so the Thomas-Fermi model may govern the screening and coexist with ionic effects (Fig. 3.4b); the screening is governed by electrons instead of ions. In a thermodynamic context, this means that charge equilibrium is governed by the statistics of electron occupation that rules the energy distribution of the molecular ensemble. The latter occupation thus predominates over the occupation of ionic states.

This theoretical analysis is experimentally reinforced using impedance-derived methods within an unprecedented resolution typified by features presented in self-assembled monolayers (SAM). An example of this are azurin-based molecular films, as shown in Fig. 3.4.

Azurin is a protein containing redox centres. When azurin is incorporated into a gold electrode modified with a self-assembled monolayer (inset of Fig. 3.4a), it clearly exhibits redox activity, as indicated by the potential-current (cyclic voltammetry) scans depicted in Fig. 3.4a. The use of a gold contact as the electrode is highly favourable from the experimental standpoint, since the self-assembly of molecular films on it is very well known and widely used [37]. In the absence of azurin (black curve), only non-Faradaic electrochemical current is attained, whereas in azurin-based films a highly reversible redox response is also visible (red curve).

Indeed, in Fig. 3.4a, Faradaic (within Thomas–Fermi screening) and non-Faradaic (within Debye screening) responses in electroactive monolayers are clearly separated

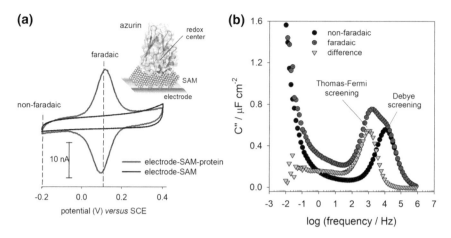

Fig. 3.4 **a** Comparison of cyclic voltammetry patterns of an electrode with (red line) and without (black line) azurin protein assembled on a non-electroactive supportive monolayer. **b** Bode plots demonstrating that non-Faradaic (black) and Faradaic (red) responses are obtained and can be spectroscopically manipulated. *Reprinted (adapted) with permission from Paulo R. Bueno, Giulia Mizzon, Jason J. Davis; Capacitance spectroscopy: A versatile approach to resolving the redox density of states and kinetics in redox-active self-assembled monolayers, The Journal of Physical Chemistry B. Copyright (2012) American Chemical Society*

using impedance-derived capacitive methods [38]. This figure (Fig. 3.4a) compares cyclic voltammetry patterns of an electrode, with (red line) and without (black line) the protein azurin, assembled on a non-electroactive supportive monolayer, using Bode plots (Fig. 3.4b), demonstrating that non-Faradaic (black) and Faradaic (red) responses are separable.

Following spectroscopic principles, a non-Faradaic response can be used as a background signal to "purify" the Faradaic response (green curve), as indicated by the analysis of $k = G/C_\mu^0$.

When impedance measurements are taken at -200 mV (indicated in Fig. 3.4a by a dashed blue line, a non-Faradaic potential region), only a relaxation peak is noticeable in the Bode plots in Fig. 3.4b (black circles), whereas two peaks are distinguishable (red circles) when measurements are taken at the Fermi energy level of the junction, corresponding to measurements made at the formal potential (indicated by the dashed blue line, Faradaic potential region, in Fig. 3.4a). By using spectroscopic concepts, the black circles, the non-Faradaic impedance-derived response can be subtracted from the red, the latter containing both non-Faradaic and Faradaic responses.

Figure 3.4b shows how an applied sinusoidal perturbation resolves Faradaic and non-Faradaic capacitive effects. The responses are based on the same physical principle as that introduced in Eq. (1.15), but their magnitude is governed by different electrical field screening phenomena. The subtracted response is shown in Fig. 3.4b in the form of green triangles, in which only the Faradaic response remains after the subtraction.

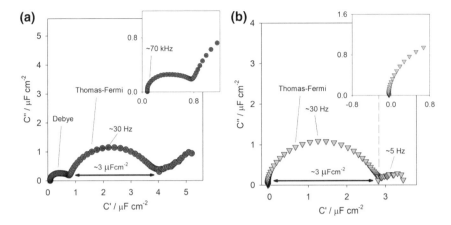

Fig. 3.5   a Nyquist diagrams illustrating Debye and Thomas–Fermi screenings contributions in molecular films containing azurin attached on a SAM-modified gold electrode. *Reprinted (adapted) with permission from Paulo R. Bueno, Giulia Mizzon, Jason J. Davis; Capacitance spectroscopy: A versatile approach to resolving the redox density of states and kinetics in redox-active self-assembled monolayers, The Journal of Physical Chemistry B. Copyright (2012) American Chemical Society*

The spectroscopic procedure illustrated in Fig. 3.4b is known as electrochemical capacitance spectroscopy [38], because it allows Faradaic processes to be resolved spectroscopically, i.e. those uniquely associated with Thomas–Fermi screening. The latter, which predominates in molecular electronics measurements, is introduced here in order to demonstrate how molecular electronics can be seen from electrochemical measurements with suitable and clear resolution [39].

The Nyquist diagrams in Fig. 3.5a illustrate Debye and Thomas–Fermi screenings contributions in molecular films containing azurin attached on a SAM-modified gold electrode (see inset in Fig. 3.4a). The measurements were taken at 90 mV versus a SCE reference potential, corresponding to the Faradaic region indicated in Fig. 3.4a. Thus, Fig. 3.5b is the same as Fig. 3.5a, but after subtraction of the non-Faradaic response. Figures 3.5a, b demonstrate that Debye and Thomas–Fermi screenings resolved using the electrochemical impedance spectroscopic approach in different ways are in agreement with Eq. (1.15).

In other words, the Bode plot shown in Fig. 3.4b, in particular, corresponds to the imaginary component $(C'' = Im[C_{\bar{\mu}}(\omega)])$ of the complex capacitance [as stated in Eq. (3.1)] plotted as a function of perturbing frequencies. Hence, the meaning of the peak depicted in the diagrams of Fig. 3.4b is directly tied to the dissipative energy of charging $C_{\bar{\mu}}^{0}$ multiplicities. The frequency of the peak corresponds to $k = G/C_{\bar{\mu}}^{0}$ in either Thomas–Fermi or Debye screening. Additionally, the spectroscopic analysis that allows for separation of Debye and Thomas–Fermi screenings is demonstrated by the Nyquist diagrams in Fig. 3.5.

Alkanes, which consist of saturated C–C bonds terminated by linkers, serve as a prototype system for experimental studies and theoretical calculations [37–43].

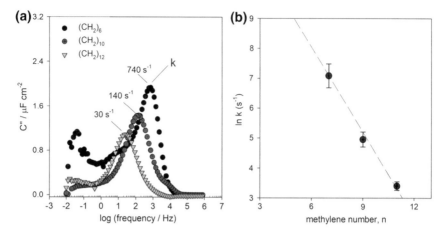

Fig. 3.6 a A comparative analysis of azurin films across three different supporting thiol layer thicknesses (hexanethiol, decanethiol and dodecanethiol) clearly shows the electron transfer rate. **b** The β (decay constant) is about 0.9 per methylene unit. *Reprinted (adapted) with permission from Paulo R. Bueno, Giulia Mizzon, Jason J. Davis; Capacitance spectroscopy: A versatile approach to resolving the redox density of states and kinetics in redox-active self-assembled monolayers, The Journal of Physical Chemistry B. Copyright (2012) American Chemical Society*

Some experiments have focused on measuring G and have found that the G of alkanes decreases exponentially with molecular length (L), such that $G = Ae^{-\beta L}$, where A is a constant and β is a decay constant varying from ~0.7 to 0.9 Angstroms^{-1}.

In general, this exponential decay allied to temperature independence [44, 45] (see Figs. 1.7c and 3.6) is suggestive of a model of electron tunnelling as the conductive mechanism for saturated chains. In varying the length of alkane-supporting layers for azurin (Fig. 3.4a, inset), impedance-derived measurements (see Fig. 3.6b) confirm that k decreases exponentially accompanying $G = kC_{\bar{\mu}}$, in agreement with complementary experiments [37, 7, 45, 46]. Comparable electron conductance or transfer rate decay trends have also been found in peptide chains, using other methods [42].

Molecular surface coverage (Γ) is a component of the magnitude of capacitance according to Eq. (3.9), affecting both Faradaic and non-Faradaic activities in molecular redox-tagged alkanes [47–50] and peptides [6, 51–53]. This is also confirmed for peptides in Fig. 3.7, according to Nyquist and Bode capacitive spectra. The overall $C_{\bar{\mu}}^0$ changes from ~10 to ~110 μF cm^{-2} [see insets of (a) and (c) in Fig. 3.7] for measurements taken in Faradaic and non-Faradaic potential regions.

The molecular coverage of electroactive peptides, calculated using a previously demonstrated equation $\Gamma = C_{\bar{\mu}}^0(4k_BT)/e^2$, is 80 pmol cm^{-2}, which is comparable to theoretically predicted values of high density packed electroactive monolayers [40]. As expected, the latter molecular coverage is higher than the 1.6 pmol cm^{-2} achieved by azurin coverage [38]. Azurin is large, so the density of its redox states must be lower than that of peptide or alkane layers, which explains the differences observed here (e.g. see Fig. 3.5 compared with that of Fig. 3.7).

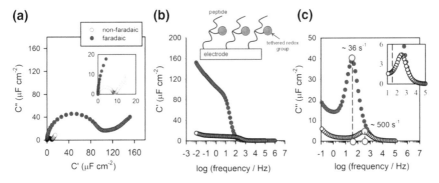

Fig. 3.7 a Nyquist capacitive diagrams demonstrating differences between electrochemical capacitance values obtained in non-Faradaic and Faradaic potential regions of a molecular redox-tagged self-assembled peptide monolayer, as illustrated schematically in the inset of (**b**). Alternatively, the same information is shown by Bode plots in (**b**) using real and (**c**) imaginary components of the complex electrochemical [$C_{\bar{\mu}}(\omega)$, Eq. (3.1)], respectively. The insets in (**a**) and (**c**) are magnifications of non-Faradaic (Debye) contributions at high frequencies of the capacitive spectra. *Reprinted (adapted) with permission from Julia P. Piccoli, Adriano Santos, Norival A. Santos-Filho, Esteban N. Lorenzón, Eduardo M. Cilli and Paulo R. Bueno; The Self-Assembly of Redox Active Peptides: Synthesis and Electrochemical Capacitive Behaviour, Biopolymer Peptide Science. Copyright (2016) Wiley*

Due to quantum effects, however, comparisons between the Γ of different molecular systems must be considered cautiously. The quantum effects associated with $C_{\bar{\mu}}^{0}$ encompass the electronic structure, which varies as a function of the external environment of the electrolyte potential; therefore, comparisons must consider the environment in which the measurements are taken. Furthermore, $\Gamma = C_{\bar{\mu}}^{0}(4k_{B}T)/e^{2}$ comprises contributions only from a single energy level of the electronic DOS enclosed in the molecular junction, precisely the Fermi level. Implications involving this limitation of the analysis will be discussed in the next section.

In summary, non-Faradaic and Faradaic phenomena, as referred to in electrochemistry, share a common theoretical ground. Differences are associated solely with unambiguous Debye or Thomas–Fermi screening governance.

3.6 Effect of the Solvent Environment and Spread of Electronic Density of States

The semiclassical electron transfer theory is supported by the transition state theory, in which electron transfer occurs in the form of $k = k_{0}\exp[-E^{*}/k_{B}T]$ and where k_{0} is the maximum rate constant and E^{*} is the activation energy. This energy generally comes from two contributing sources, one originating from the mechanical elastic distortion of internal force fields and the other from charge injections [54, 55] that occur in the presence of an external electrostatic potential accounted for in C_{ext} (see Fig. 1.6), which involves the role of the solvent's polarisation properties [56].

Neither of these energy components (contributing to E^*), alone or combined, are adequate to describe the quantum electrochemical RC model. The reason for this is that, in the Debye–Hückel theory, the fluctuations of the screened Coulomb potential operate on $\phi(r) = (q/r)e^{\kappa r}$, where r is the spatial position, q is the charge and κ is the inverse of the Debye length within a Boltzmann approximation.

However, following classical approximations, κ is unsuitable according to previous reasoning (see Chap. 1, Sect. 1.8) that validates $\kappa = \left(C_{\bar{\mu}}^0 / \varepsilon_r \varepsilon_0 \right)^{1/2}$ instead of $\kappa = \left[(2e^2 \mathcal{N}) / (\varepsilon_0 \varepsilon_r k_B T) \right]^{1/2}$, although $\kappa \propto (1/\varepsilon_0 \varepsilon_r)^{1/2}$ in both situations. In fact, the meaning of capacitance in one type of electric field screening (Debye) is totally distinct from the other (Thomas–Fermi) [57].

Accordingly, note that although semiclassical electron transfer theory is generally adequate to predict the impact of the dielectric properties of the environment on the electron transfer rate [56] because of Debye *versus* Thomas–Fermi electric field screening, it fails to predict how solvent reorganises and then governs the HOMO and LUMO differences (whose inverse is directly associated with $C_{\bar{\mu}}^0$, as shown in Chap. 2) [7].

In terms of heterogeneous electron transfer, the way in which states between HOMO and LUMO levels couple with an electrode is regulated by $\kappa = \left(C_{\bar{\mu}}^0 / \varepsilon_r \varepsilon_0 \right)^{1/2}$ [3]. An in-depth evaluation has yet to be done of the effects of $\kappa = \left(C_{\bar{\mu}}^0 / \varepsilon_r \varepsilon_0 \right)^{1/2}$ on the reorganisation energy, which is a key parameter in electron transfer theory.

To exemplify briefly, as a matter of preliminary discussion of this assumption, Fig. 3.8b shows, in an 11-ferrocenyl-undecanethiol monolayer assembled on gold electrodes, how the Gaussian shape experimentally obtained from the electronic DOS spreads through the dielectric environment, in line with Eq. (3.10) and with $\kappa = \left(C_{\bar{\mu}}^0 / \varepsilon_r \varepsilon_0 \right)^{1/2}$.

Note that the electronic DOS associated with Eq. (1.15) (right hand) is not the nuclear Franck–Condon weighted DOS (the left-hand one has been semiclassically developed) [31, 58]. Both DOS functions affect the electron transfer rate; however, from a detailed quantum statistical mechanics point of view, it is not yet completely understood how they are fundamentally correlated, so a more in-depth evaluation is still necessary.

Moreover, the quantum effects are remarkable and the differences in the electronic structure of the molecular film (Fig. 3.8) as a function of variations in the dielectric environment cannot be explained classically or semiclassically. The electronic structure is wave or electron density dependent, and the spatial position of the atoms alone cannot determine the physical properties of the system.

Figure 3.8 illustrates the impact of the static dielectric constant of the solvent (water $\varepsilon_s \sim 80$ [yellow], acetonitrile $\varepsilon_s \sim 40$ [green] and dichloromethane $\varepsilon_s \sim 10$ [red]) on experimentally resolved k and $C_{\bar{\mu}}^0$ for a 11-ferrocenyl-undecanethiol film showing mean values of three independent measurements. $C_{\bar{\mu}}^0$ is intrinsically associated with the dielectric constant, according to $C_{\bar{\mu}}^0 = \varepsilon_r \varepsilon_0 \kappa^2$.

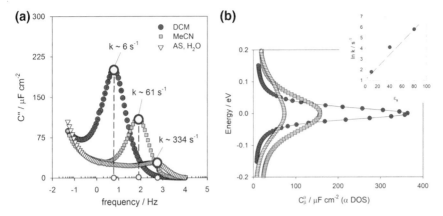

Fig. 3.8 Electron transfer rate, k, is affected by the solvent environment, as shown in (**a**) and in the inset. **b** Shows the dispersion of C_μ^0 and DOS Gaussian function [as predicted by Eq. (3.10)] as a consequence of the effects of the dielectric environment on the quantum RC ensemble. *Reprinted (adapted) with permission from Paulo R. Bueno and Jason J. Davis; Measuring Quantum Capacitance in Energetically Addressable Molecular Layers, Analytical Chemistry. Copyright (2014) American Chemical Society*

Note that, predictably, the integrated charge, i.e. the value of total DOS (given by the area obtained by integrating the Gaussian curves), is constant. The electron transfer rate, k, is affected, as shown in (a) and in the inset. Figure 3.8b shows the dispersion of C_μ^0 and DOS Gaussian function [as predicted by Eq. (3.10)] as a consequence of the effects of the dielectric environment on the quantum RC ensemble. The inset in Fig. 3.8b demonstrates the linear relationship of logarithmic k and ε_s, ultimately related to Marcus theory, but with reservations. In Fig. 3.8b, note that the Fermi level associated with the formal electrochemical potential of the electrode was zeroed simply for convenience, but it varies according to the solvent environment.

For instance, naively misinterpreting Eq. (3.9) for this particular situation and following the previous reasoning, one would be able to calculate the molecular coverage as $\Gamma = C_\mu^0 (4RT)/e^2$ at the formal potential of the junction. C_μ^0 maximises for the latter, as shown by the peaks in Fig. 3.8b, at ~380, ~170 and ~80 μF cm^{-2} for dichloromethane (DCM), acetonitrile (MeCN) and aqueous solvent (AS, H$_2$O), respectively, as inferred from the three different positions of the maximum value of C_μ^0 in Fig. 3.8b. Therefore, three different values of capacitance or molecular coverage would be obtained for the same 11-ferrocenyl-undecanethiol molecular film when placed in different solvents.

The Gaussian DOS responses in different dielectric environments are reversible, so the molecular coverage and the film's atomic structure remain "steady" (there is neither loss of ferrocene electrochemical centres to the environment nor a decrease in electrochemical activity). Why, then, does ferrocene molecular coverage vary in different dielectric environments? This is not an inaccuracy resulting from theory, but is based on our common classical interpretive understanding of quantum effects (sometimes not even recognised as quantum).

In fact, what is incorrect is our classical thinking, because now it is Eq. (3.10) that rules, not Eq. (3.9) from which $\Gamma = C_\mu^0 (4RT)/e^2$ was derived. By integrating all the energy levels, the total electronic DOS is obtained and is thus constant as $1.37 \pm 0.4 \times 10^{14}$ states cm^{-2}, demonstrating that the correct interpretation is allowed by the theory, i.e. the electron density remains constant throughout all levels of charging energy.

The impact of the dielectric environment is distributed over the electronic structure of the junction. This structure only reorganises its electron density [5] differently, following the polarisation associated with C_{ext}, although electron density at the molecular junction remains constant regardless of the dielectric environment. Note that the solvent effects reported here are in contrast with electrochemical DOS occupancy, when molecular targets are captured by a receptive layer concomitantly attained at the surface; see Fig. 3.9b and related discussion below.

In summary, the intrinsic existence of an electronic DOS beyond the nuclear Franck–Condon weighted DOS per se is not predicted by semiclassical electron transfer rate theory. Hence, the effect of solvent on the electronic DOS cannot be determined based on this theory, but by the theory of mesoscopic systems that underpins the quantum RC circuit.

3.7   Energy Transducer and Sensing

We will now explain how previous concepts can be used to design molecular diagnostic assays, using a unique example of quantum RC electrochemistry applied to sensing devices. If an interface containing an ensemble of individual electrochemical capacitors (representing the capacitance associated with redox-active molecules) is bound with receptive biological receptors, a sensorial interface is constructed (Fig. 3.9a), given that the electron occupancy of the ensemble is sensitive to the biological binding event (Fig. 3.9b). Such a biosensing interface has proven useful in developing molecular diagnostic devices using redox-active alkanes [47–50, 59–61] and redox-tagged peptide monolayers [6, 51–53] or even graphene layers [32].

The variation in energy of the interface according to the number of electrons occupying the electrochemical capacitor ensemble is modulated by a chemical potential difference, such that $dE/d\langle\mathcal{N}\rangle = \langle\mathcal{N}\rangle(e^2/C_\mu^0)$, where $\left(e^2/C_\mu^0\right)$ is the energy required to charge the capacitors individually (as demonstrated previously). Such a change in interfacial energy can be appropriately used as a transducer signal of the biological binding event.

Figure 3.9a depicts an example of a molecular redox-tagged interface. Two important characteristics should be noted. The first has to do with the difference between the effects of the solvent (previously shown in Fig. 3.8), while the second pertains to the effect of the environment due to the occupancy of receptor sites at the interface.

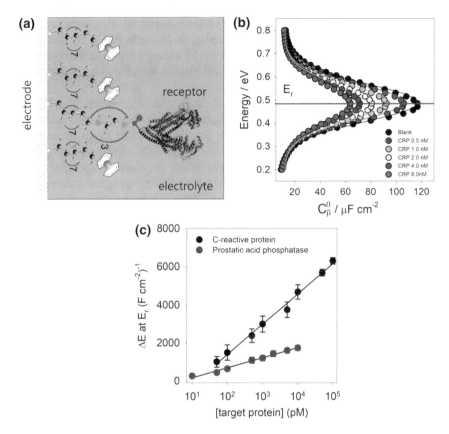

Fig. 3.9 **a** Illustration of electroactive species bound to receptive biological receptors, forming an electroactive sensorial (field-effect sensitive) interface that can be used in molecular diagnostics. **b** The responsiveness of electrochemical DOS to molecular recognition within a mixed redox-switchable and antibody-constrained film, as shown in (**a**). In (**c**), $\Delta E = (e^2/C_{\bar{\mu}}^0)$ corresponds to the amount of energy per electrons modified in the electrode junction as a consequence of binding equilibrium during recognition of the target as a function of its concentration in the electrolyte environment. *Reprinted (adapted) with permission from Paulo R. Bueno and Flavio C. B. Fernandes and Jason J. Davis; Quantum Capacitance as a Reagentless Molecular Sensing Element, Nanoscale. Copyright (2017) Royal Society of Chemistry*

As can be seen in Fig. 3.9b, a change in the local dielectric constant translates into a resolved DOS energy redistribution *without a change in total state occupancy* $[\Delta \langle \mathcal{N} \rangle_t = (1/e^2) \int_{-\infty}^{+\infty} C_{\bar{\mu}}(V) dV$, as expressed [8] by the total area of the normal DOS distribution function, following the rationale stated in Eq. (3.10)]. However, a molecular binding event (such as that occurring at a neighbouring receptive site) triggers a resolved change in local and total DOS occupation (Fig. 3.9b).

Figure 3.9a demonstrates how electroactive species bound to receptive biological receptors form an electroactive sensorial interface (field-effect sensitive) that can be used as a molecular transistor for molecular diagnostics. Figure 3.9b demonstrates the responsiveness of electrochemical DOS to molecular recognition within a mixed redox-switchable and antibody-constrained film, as shown in Fig. 3.9a. Moreover, Fig. 3.9b also shows the invariance in both Fermi energy of the junction (~0.49 V vs. Ag|AgCl chemical reference) and electronic states dispersion as a function of the concentration of C-reactive protein target.

Figure 3.9b illustrates the experimental response of an electroactive film, like the one shown in Fig. 3.9a, with integrated biological receptors. When a target biomarker is recruited, the occupation of the electronic states changes and the experimentally measured change in interfacial energy is transduced into DOS occupancy. In Fig. 3.9c, $\Delta E = (e^2/C_{\mu}^0)$ corresponds to the amount of energy per electrons modified in the electrode junction as a consequence of binding equilibrium during recognition of the target as a function of its concentration in the electrolytic environment.

We can try to distinguish between how the Gaussian DOS in Figs. 3.8b and 3.9b is environmentally triggered by examining the DOS shape (which follows a normal distribution) and its associated energy spread as expressed through $S[d\langle\mathcal{N}\rangle/dE] = \ln \sigma_g\sqrt{2\pi e_n}$ (the entropy function of the normal DOS distribution), where e_n is the base of natural logarithms and σ_g is the standard deviation. When there is a dielectric change in the environment (Fig. 3.8b), there is a clearly resolved change in dispersion (Fig. 3.8b) and Fermi-level displacement (not shown). Nonetheless, the effects of a local binding event (Fig. 3.9b) are different in that there is an associated change in chemical potential, resolvable through DOS occupation as measured by C_{μ}^0 *without* any change in electronic dispersion (Fig. 3.9b) or on the Fermi level.

In other words, the electronic DOS presented by a redox molecular film responds to a neighbouring molecular recognition (Fig. 3.9b), such that the resolved electronic charge, $\Delta\langle\mathcal{N}\rangle$, is proportional to the target/analytical concentration (Fig. 3.9c; the target here being C-reactive protein, CRP, and prostatic acid phosphatase, PAP). The energy-related signal pertains directly to the electron occupation of quantised states. Sensorial interfaces can thus be designed based on the response of a confined and resolved electronic density of states to target binding and the associated change at the interfacial chemical potential [32]. This concept has been demonstrated with a number of clinically important markers, representing a new potent and ultrasensitive molecular detection method [6, 47–53, 62] (femto- to picomolar), which allows for an energy transducer principle capable of quantifying, in a single step and in a reagentless manner, markers within biological fluid [32, 59, 61, 63, 64, 47–49, 51].

In contrast to field-effect transistor devices, in which the modulation of conductance (flow of electrons or holes) is controlled by electrostatic potential variation (in the gated environment) of the charge density in the conduction channel, the methodology of a quantum electrochemical field-effect device, as introduced above, is based on the detection of the local change in charge density, or the so-called field effect [see considerations following Eq. (3.7)]. This effect characterises the recognition event between a target molecule and the surface receptor, but the sensitivity is driven by the density of the state of the quantum capacitance rather than by the flow of DC current through the channel, as normally assumed in three-terminal measurement experiments (see Fig. 3.2a) in traditional FETs [27, 28].

It has been shown, in this section, that a surface-confined and electronically addressable DOS (comprising either redox-switchable centres or appropriately immobilised mesoscopic units [32]) contributes to a readily resolved electrochemical capacitance that is strongly dependent on quantum effects. This charging relies on the occupancy of quantised states and responds sensitively to changes in local electrochemical potential caused by the molecular occupation of an integrated receptive layer. By introducing receptors appropriately, this entirely reagentless sensing becomes highly specific and remarkably sensitive.

In summary, the transduction mechanism of molecular switches operates in a manner analogous to FET devices, but in a markedly more experimentally accessible and chemically flexible manner, allowing commercially viable molecular quantum devices to be fabricated. It has been shown that [32] the combination of this approach with standard microfluidic and microfabrication methods would offer a great deal in the near future to both diagnostics and chemical sensing in general. In the next section, we will discuss how the theory aligns with electrochemical capacitance to calculate the conductivity of DNA molecules.

3.8 Quantum Conductance of DNA Wires

Information in a protein chain can be transmitted over great distances through charge transport. There are two types of electron transport mechanisms: (i) electron tunnelling or superexchange and (ii) electron hopping [65, 42]. Tunnelling is a coherent electron transfer process because bridge levels are not occupied due to the high excitation gap between the donor states and bridge states, whereas hopping is an incoherent process in which an electron occupies the levels of the bridging medium during its transport from the donor to the acceptor.

In a hopping mechanism, the molecular bridge not only links the electron donor to the acceptor electronically but also, in the case of peptides and similar molecular structure, involves its amino acids in oxidation and reduction, thus providing relay stations or stepping stones for the transport of electrons. The involvement of particular amino acids as redox-active intermediates renders the distant electron transfer a multistep tunnelling process (hopping process) whose kinetics are faster than that of a long single-step electron transfer between donor and acceptor.

We are presently interested in the observation that "wires" made of molecules can connect electrochemical capacitive resonant centres (for instance ferrocene) to the electrode (see Figs. 1.1c and 3.10a); these molecular wires (sometimes called molecular nanowires or chains) act as a bridge over which electrons can be transmitted or transported along the molecular electronic structure by hopping, enabling transport over large distances. These molecular chains that conduct electric current are proposed as building blocks for future nanoelectronic devices, and a widely accepted example of this is DNA. DNA strands are used as nanowires (with a length of ~7 nm), and their propensity to transmit or transport charge can be evaluated from the dynamics of a quantum RC electrochemical circuit, as introduced earlier, simply

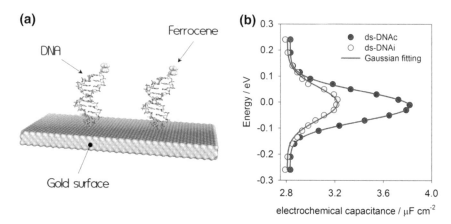

Fig. 3.10 **a** Illustration of the covalent attachment and electronic coupling of redox-tagged (ferrocene) double-stranded DNA (about 7 nm in length) onto a gold electrode. **b** Keeping the environment (electrolyte) stationary and length constant, an electronic DOS can be constructed for different nucleobase sequences in which charge transport is expected to be different. *Reprinted (adapted) with permission from William W. C. Ribeiro, Luís M. Gonçalves, Susana Liébana, Maria I. Pividori Paulo R. Bueno; Molecular Conductance of Double-stranded DNA Evaluated by Electrochemical Capacitance Spectroscopy, Nanoscale. Copyright (2016) Royal Society of Chemistry*

by applying the relationship [14] $C_\mu^0 = G/k$. Regardless of whether the process is tunnelling or hopping, the relationship $C_\mu^0 = G/k$ may be present and the influence of the electronic structure implicitly achieved by C_μ^0 and the contained DOS shape.

Figure 3.10a illustrates the covalent attachment and electronic coupling of redox-tagged (ferrocene) double-stranded DNA (about 7 nm in length) onto a gold electrode. Figure 3.10b demonstrates that by keeping the environment (electrolyte) stationary and the length constant, an electronic DOS can be constructed for different nucleobase sequences governed by different charge transport mechanisms.

The hypothesis is that charge transport in dsDNA is a function of the distance between the GC base pair, with the electronic transfer rate decreasing by about one order of magnitude with each intervening AT base pair, and where the electrical response decreases exponentially in response to increasing amounts of adenine and thymine. This is why AT sites are known to act as conduction barriers. Charge flows through poly(dG)-poly(dC) (dsDNAc) are therefore expected to be comparatively easier than through poly(dA)-poly(dT) (dsDNAi) [66–69].

The simplest analysis can be performed at the Fermi level and low frequencies at which $G = kC_\mu^0 = k\left(e^2\Gamma/4k_BT\right)$ holds. Based on the latter relationship, the conductances of DNA strands (made of different nucleobase sequences) were found to be measurable [70]. The conductances obtained for dsDNAc and dsDNAi were, respectively, 4.6 ± 0.4 S m^{-1} and 0.48 ± 0.05 S m^{-1} (in resistivity values, ~0.2 Ω m and ~2.1 Ω m) [70]. The conductance values measured by impedance spectroscopy were consistent with those obtained by other methodologies such as scanning tunnelling microscopy (STM) [63, 64].

Although only DNA conductivity (in different chemically designed double strands) has so far been tested, the theory and methodology outlined here is simple and also suitable for determining the conductivity of other molecular wires.

It is also important to consider the comparison between molecular electronics (through direct conductance measurements) and electrochemistry (indirectly inferred from measured charge transfer rates) introduced recently, in which a power-law relationship [71] is empirically demonstrated. This empirical power-law existing between molecular conductance and the electron transfer rate was shown to occur in different molecular wires [71]. It has recently been demonstrated that $G = kC_{\mu}^0$ theoretically elucidates power-law dependence [39].

In summary, in this section we demonstrated that the relationship of the quantum of conductance and the electrochemical capacitance, which is $G = kC_{\mu}^0$, can be used experimentally to determine the conductance of molecular wires. In the next section, we will demonstrate that the relationship, $k = G/C_{\mu}^0$, also explains supercapacitance and pseudocapacitance phenomena that usually drive electrochemical devices used for energy storage.

3.9 Supercapacitance or Pseudocapacitance?

Since $C_{\mu}^0 = G/k$ is based on first-principle quantum mechanics, its generality is expected to drive us beyond molecular conductance or biosensing applications. This statement is supported here by demonstrating the applicability of C_{μ}^0 to elucidate capacitance phenomena in graphene compounds.

Graphene is one of the most intriguing molecular compounds and, owing to its unique and extraordinary electronic properties, it has been speculated that graphene can carry a supercurrent (bipolar current) [72]. A fact that has been ignored or neglected is that the outstanding electron transport properties of graphene are intrinsically associated with a large capacitance, which, from a classical mechanics point of view, is antagonistic to electron transport.

Accordingly, despite numerous reviews on the properties and applications of graphene [73–75], only a few studies have correlated its electronic conductive characteristics with its outstanding energy storage capacity [76]. The origin of the molecular supercapacitance of electrochemically modified mesoscopic graphene layers and its behaviour when embedded in an electrolyte was only recently elucidated through the use of the quantum RC electrochemical model introduced earlier herein [77].

Figure 3.11a shows typical Bode plots, while Fig. 3.11b shows typical Nyquist diagrams obtained for carbonaceous materials. The response of a glassy carbon electrode (GCE) is used as a reference value for the capacitances obtained when GCE is used as a substrate for graphene oxide (GO) and reduced graphene oxide (RGO) layers. The inset in Fig. 3.11b is zoomed in at the high-frequency region for a closer view of the differences. These diagrams also show the RC time constant associated with charging of capacitive states, which are proportional to $R_q C_{\mu}^0$.

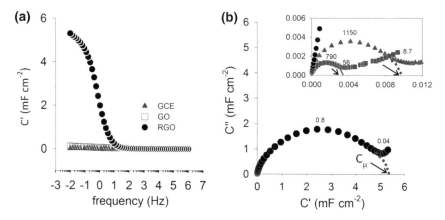

Fig. 3.11 Typical **a** Bode and **b** Nyquist diagrams obtained for GCE (glassy carbon electrode, as a reference for capacitance values and used as substrate) (black up-pointing triangle), GO (graphene oxide) (black square) and RGO (reduced graphene oxide) (black circle) in aqueous electrolyte of 0.05 M phosphate-buffered saline. The inset in **b** is zoomed in at the high-frequency region. *Reprinted (adapted) with permission from Fabiana A. Gutierrez, Flavio C. B. Fernandes, Gustavo A. Rivas and Paulo R. Bueno; Mesoscopic behaviour of multi-layered graphene: the meaning of supercapacitance revisited, Physical Chemistry and Chemical Physics. Copyright (2017) Royal Society of Chemistry*

In the case of RGO, the contribution of the quantum RC time constant explains the abrupt increase in capacitance values by the electrochemical reduction of graphene oxide, which is equivalent to chemical doping. GCE and GO have comparable capacitances, but the capacitance of RGO (after reduction of GO in the same electrode) is more than three orders of magnitude higher. The explanation for this has to do with changes in the density of states of GO when it is reduced, which clearly demonstrated the quantum effects that impact the RC time constant at the interface. The quantum accessible states in RGO contribute to additional charging states and concomitantly increase the entire electrochemical RC time constant.

Indeed, Fig. 3.11 indicates that the value of capacitance obtained for reduced graphene oxide is very high, undeniably within mF cm^{-2} (~5.1 mF cm^{-2}), which is about 1000-fold higher than that obtained for graphene oxide (typically between 2 and 10 μF cm^{-2}). The associated increase in the electroactive area is approximately 25-fold, which is much lower than the increase observed in capacitance. Accordingly, this increase in capacitance cannot be attributed solely to an increase in the specific area and geometric capacitive effects associated with double-layer phenomena. It should be emphasised that the only difference between reduced graphene oxide and graphene oxide per se is associated with the electrochemical reduction of graphene oxide.

This experimental observation then enabled a revisit to the origin of graphene supercapacitors, where the electronic density of states of graphene was found to be crucial in understanding its supercapacitance phenomena, which cannot be explained solely by a double-layer model or by non-Faradaic interfacial charging effects. The higher accessible electronic DOS observed within reduced graphene oxide was associated with the dominance of a quantum RC electrochemical effect [77] (see Fig. 3.11). The capacitance of reduced graphene oxide is associated with charging electronic states; hence, a mesoscopic effect is predominant in behaviour predicted by the quantum electrochemical RC circuit and by Faradaic processes controlled by Thomas–Fermi screening, as discussed earlier in Sect. 3.5.

In summary, the RC characteristics associated with mesoscopic electron relaxation, where R_q and C_μ^0 are interdependent, demonstrate that the higher the conductance (lower resistance) associated with C_μ^0 accessible states, the higher the supercapacitive effect in graphene compounds. This information is obviously useful for designing carbonaceous materials and interfaces for supercapacitor applications.

3.10 Pseudocapacitance of Semiconductor Materials and Sensing of Low-Molecular-Weight Target Molecules

Similarly to graphene, as discussed in the preceding section, electrodes made of highly ordered n-type TiO_2 semiconducting nanotubes (TNTs) have emerged as a viable alternative of energy storage devices, owing to the easy modification of Ti metal plates via electrochemical anodisation (see Fig. 3.12) [70]. This yields TNT-modified electrodes of suitable electronic conductivity and high surface area. These electrodes acquire particular properties when self-doped by means of subsequent cathodic polarisation, and are therefore known as self-doped TNT (SD-TNT).

TNT-based electrodes have been successfully applied for the photo(electro)catalytic oxidation of contaminants, dye-sensitised solar cells [78], water splitting [31] and photoelectrochemical sensors [79]. On the other hand, complementary SD-TNTs have been used in supercapacitor devices [80–84], electrochemical oxidative cells [82, 84–86], cleansing electrochemical systems [86, 5, 57], photoanodes for water splitting cells [33], and as anodes in lithium ion batteries [58].

TNT-based electrodes are cathodically polarised under a negative potential, usually lower than -1.0 V, which generates oxygen vacancies associated with Ti_4O_7 clusters and Ti^{3+}/Ti^{4+} mixed valences [80]. This electrochemical doping process involves charge compensation through the intercalation of protons [$Ti^{4+}O_2 + e^- + H^+ \leftrightarrows Ti^{3+}(O)(OH)$]. It is estimated that up to about 1% of Ti^{4+} can be reduced to Ti^{3+} [80], which induces a metallic-like behaviour [83] within a concomitant, intrinsic and proportionally higher capacitive behaviour (in the order of mF cm^{-2}).

In this section, we will demonstrate that TiO_2 nanotube electrodes, when self-doped (SD-TNT), provide a concomitant increase in conductance and capacitance. Accordingly, increases of 100-fold (from 19.2 ± 0.1 μF cm^{-2} to 1.9 ± 0.1 mFcm^{-2}) and twofold (from ~6.2 to ~14.4 mS cm^{-2}) in capacitance and conductance can be achieved, respectively.

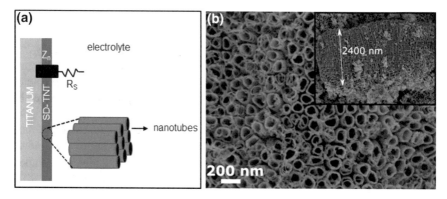

Fig. 3.12 **a** Schematic representation of a self-doped TiO$_2$(SD-TNT) nanotube grown on a titanium metal plate embedded in an electrolyte. **b** SEM micrograph of self-doped TiO$_2$ as synthesised in the present work. *Reprinted (adapted) with permission from Felipe F. Hudari, Guilherme G. Bessegato, Flávio C. Bedatty Fernandes, Maria V. B. Zanoni and Paulo R. Bueno;* Reagentless Detection of Low-Molecular-Weight Triamterene Using Self-Doped TiO$_2$ Nanotubes, *Analytical Chemistry. Copyright (2018) American Chemical Society*

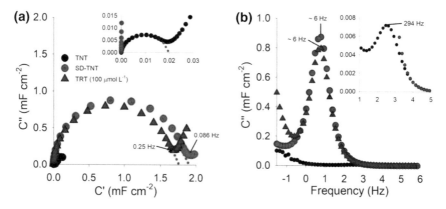

Fig. 3.13 **a** Nyquist diagrams of TNT, SD-TNT and SD-TNT in the presence of the TRT target. **b** Bode plots of the imaginary capacitance (C'') component. *Reprinted (adapted) with permission from Felipe F. Hudari, Guilherme G. Bessegato, Flávio C. Bedatty Fernandes, Maria V. B. Zanoni and Paulo R. Bueno;* Reagentless Detection of Low-Molecular-Weight Triamterene Using Self-Doped TiO$_2$ Nanotubes, *Analytical Chemistry. Copyright (2018) American Chemical Society*

The resistance of the nanotube electrodes was determined based on the Nyquist diagrams shown in Fig. 3.13, similarly as was done in the previous section for graphene. The resistance decreased significantly from ~161 Ω cm^2 (TNT) to ~70 Ω cm^2 (SD-TNT). This corresponds to an increase in conductance from ~6.2 mS cm^{-2} (TNT) to ~14.4 mS cm^{-2} (SD-TNT).

Also interesting to note is that the ratio of conductance to capacitance can be used to detect and quantify, in a reagentless manner and specifically, the low-molecular-weight triamterene (TRT) [87] diuretic by designing an appropriate doping level of TiO_2 nanotubes. Accordingly, Fig. 3.13a shows the Nyquist capacitance diagrams for TNT, SD-TNT and SD-TNT in the presence of the TRT target.

In Fig. 3.13, note the variation in the real and imaginary capacitance terms as a function of TRT concentrations, which is shown in the Nyquist plot. The measurements were recorded in a 0.10 mol L^{-1} B–R buffer solution (pH 2.0) in the frequency range of 1–30 MHz. $C_{\bar{\mu}}^0$ was measured at lower frequencies, and its diameter is approximately that of the semicircle depicted in the Nyquist diagrams.

Figure 3.13b shows the Bode plots for the imaginary capacitance (C'') component of the complex capacitance. Note that the lower frequency region of the spectrum is significantly changed in the presence of TRT. Note, also, that the electronic resonance frequency is not affected by the presence of TRT [87], implying that $k = G/C_{\bar{\mu}}^0$ is approximately constant. Nonetheless, there is an evident process of energy loss in the frequency range of 86–250 MHz (lowest-frequency region) associated with the presence of TRT [87].

$C_{\bar{\mu}}^0$, which was determined directly from the size of the semicircle in the Nyquist plot (Fig. 3.13), was 19.2 ± 0.1 $\mu F\ cm^{-2}$ (microfarad) for TNT and greatly increased to 1.9 ± 0.1 $mF\ cm^{-2}$ (millifarad) for SD-TNT. The conductive and capacitive values determined experimentally for TNT and SD-TNT are likely associated with Ti_4O_7 clusters [37, 39], as previously discussed.

TiO_2 structures containing different concentrations of Ti_4O_7 clusters have been confirmed to be very different in terms of their electron density of states, given that $C_{\bar{\mu}}^0 \propto d\mathcal{N}/d\bar{\mu}$ [37]. Accordingly, the pseudocapacitance herein interpreted as $C_{\bar{\mu}}^0$ is characteristically dependent on the self-doping process; the capacitance increases from microfarads (per unit of electroactive area) in TNTs to millifarads (per unit of electroactive area) in SD-TNTs (a 100-fold increase). Similar SD-TNT capacitive behaviour has been reported by different authors [81, 82], confirming that SD-TNT exhibits supercapacitive or pseudocapacitive behaviour that can be explained based on the fundamental concepts introduced in this book [80, 88].

The detection of TRT by the SD-TNT electrode can be explained by the interaction of electronic states of SD-TNT with the nitrogen atoms present in diuretic molecules. At 1.8 V, an energy alignment is possible between the "HOMO" and "LUMO" states of TRT and these states in self-doped TiO_2 semiconducting nanotubes.

The analysis presented here can evidently be extended to include mesoscopic structures (beyond graphene or TNTs) that may be coupled to macroscopic electrodes in the presence of an electrolyte.

References

1. A.J. Bard, L.R. Faulkner, *Electrochemical Methods Fundamentals and Applications*, 2nd edn. (Wiley, New York, 2000)
2. G.A. Snook, P. Kao, A.S. Best, Conducting-polymer-based supercapacitor devices and electrodes. J. Power Sources **196**(1), 1–12 (2011)
3. G.P. Wang, L. Zhang, J.J. Zhang, A review of electrode materials for electrochemical supercapacitors. Chem. Soc. Rev. **41**(2), 797–828 (2012)
4. L.L. Zhang, X.S. Zhao, Carbon-based materials as supercapacitor electrodes. Chem. Soc. Rev. **38**(9), 2520–2531 (2009)
5. D.A. Miranda, P.R. Bueno, Density functional theory and an experimentally-designed energy functional of electron density. Phys. Chem. Chem. Phys. **18**(37), 25984–25992 (2016)
6. A. Santos, J.P. Piccoli, N.A. Santos, E.M. Cilli, P.R. Bueno, Redox-tagged peptide for capacitive diagnostic assays. Biosens. Bioelectron. **68**, 281–287 (2015)
7. P.R. Bueno, G.T. Feliciano, J.J. Davis, Capacitance spectroscopy and density functional theory. Phys. Chem. Chem. Phys. **17**, 9375–9382 (2015)
8. M.S. Gudiksen, L. J. Lauhon, J. Wang, D.C. Smith, C.M. Lieber, Growth of nanowire super-lattice structures for nanoscale photonics and electronics, Nature **415**(6872), 617–620 (2002)
9. K. Ariga, J.P. Hill, Q.M. Ji, Layer-by-layer assembly as a versatile bottom-up nanofabrication technique for exploratory research and realistic application. Phys. Chem. Chem. Phys. **9**(19), 2319–2340 (2007)
10. J.N. Chen, M. Badioli, P. Alonso-Gonzalez, S. Thongrattanasiri, F. Huth, J. Osmond, M. Spasenovic, A. Centeno, A. Pesquera, P. Godignon, A.Z Elorza, N. Camara, F.J.G. de Abajo, R. Hillenbrand, F.H.L. Koppens, Optical nano-imaging of gate-tunable graphene plasmons. Nature **487**(7405), 77–81 (2012)
11. Z.H. Chen, Y.M. Lin, M.J. Rooks, P. Avouris, Graphene nano-ribbon electronics. Physica E-Low-dimensional systems and nanostructures, **40**(2), 228–232 (2007)
12. D. Grieshaber, R. MacKenzie, J. Voros, E. Reimhult, Electrochemical biosensors-sensor principles and architectures. Sensors, **8**(3), 1400–1458 (2008)
13. H.I. Hanafi, S. Tiwari, I. Khan, Fast and long retention-time nano-crystal memory. IEEE Trans. Electron Devices **43**(9), 1553–1558 (1996)
14. M.C. McAlpine, H. Ahmad, D.W. Wang, J.R. Heath, Highly ordered nanowire arrays on plastic substrates for ultrasensitive flexible chemical sensors. Nat. Mater. **6**(5), 379–384 (2007)
15. V. Singh, D. Joung, L. Zhai, S. Das, S.I. Khondaker, S. Seal, Graphene based materials: past, present and future. Prog. Mater Sci. **56**(8), 1178–1271 (2011)
16. Z.L. Wang, Functional oxide nanobelts: materials, properties and potential applications in nanosystems and biotechnology. Annu. Rev. Phys. Chem. **55**, 159–196 (2004)
17. K. Balasubramanian, M. Burghard, Biosensors based on carbon nanotubes. Anal. Bioanal. Chem. **385**(3), 452–468 (2006)
18. C. Bartic, G. Borghs, Organic thin-film transistors as transducers for (bio) analytical applications. Anal. Bioanal. Chem. **384**(2), 354–365 (2006)
19. J. Vanderspiegel, I. Lauks, P. Chan, D. Babic, The extended gate chemically sensitive field-effect transistor as multi-species microprobe. Sens. Actuators **4**(2), 291–298 (1983)
20. Y. Ohno, K. Maehashi, Y. Yamashiro, K. Matsumoto, Electrolyte-gated graphene field-effect transistors for detecting pH protein adsorption. Nano Lett. **9**(9), 3318–3322 (2009)
21. S. Rosenblatt, Y. Yaish, J. Park, J. Gore, V. Sazonova, P.L. McEuen, High performance electrolyte gated carbon nanotube transistors. Nano Lett. **2**(8), 869–872 (2002)
22. M. Berggren, A. Richter-Dahlfors, Organic bioelectronics. Adv. Mater. **19**(20), 3201–3213 (2007)
23. C.R. Bondy, S.J. Loeb, Amide based receptors for anions. Coord. Chem. Rev. **240** (1–2), 77–99 (2003)
24. R.H. Friend, R.W. Gymer, A.B. Holmes, J.H. Burroughes, R.N. Marks, C. Taliani, D.D.C. Bradley, D.A. Dos Santos, J.L. Bredas, M. Logdlund, W.R. Salaneck, Electroluminescence in conjugated polymers. Nature **397** (6715), 121–128 (1999)

25. M.; Kaempgen, C.K. Chan, J. Ma, Y. Cui, G. Gruner, Printable thin film supercapacitors using single-walled carbon nanotubes. Nano Lett. **9**(5), 1872–1876 (2009)
26. U. Lange, N.V. Roznyatouskaya, V.M. Mirsky, Conducting polymers in chemical sensors and arrays. Anal. Chim. Acta **614**(1), 1–26 (2008)
27. P. Lin, F. Yan, Organic thin-film transistors for chemical and biological sensing. Adv. Mater. **24**(1), 34–51 (2012)
28. J.T. Mabeck, G.G. Malliaras, Chemical and biological sensors based on organic thin-film transistors. Anal. Bioanal. Chem. **384**(2), 343–353 (2006)
29. I. McCulloch, M. Heeney, C. Bailey, K. Genevicius, I. Macdonald, M. Shkunov, D. Sparrowe, S. Tierney, R. Wagner, W.M. Zhang, M.L. Chabinyc, R.J. Kline, M.D. McGehee, M.F. Toney, Liquid-crystalline semiconducting polymers with high charge-carrier mobility. Nat. Mater. **5**(4), 328–333
30. J. Piccoli, R. Hein, A.H. El-Sagheer, T. Brown, E.M. Cilli, P.R. Bueno, J.J. Davis, Redox capacitive assaying of c-reactive protein at a peptide supported aptamer interface. Anal. Chem. **90**(5), 3005–3008 (2018)
31. P.R. Bueno, T.A. Benites, J.J. Davis. The mesoscopic electrochemistry of molecular junctions. Sci. Rep. **6**, 18400 (2016)
32. P.R. Bueno, F.C.B. Fernandes, J.J. Davis, Quantum capacitance as a reagentless molecular sensing element. Nanoscale (2017)
33. S. Luryi, Quantum capacitance devices. Appl. Phys. Lett. **52**, 501 (1988)
34. W.T. Yang, R.G. Parr, Hardness, softness and the fukui function in the electronic theory of metals and catalysis, in *Proceedings of the National Academy of Sciences of the United States of America*, **82** (20), 6723–6726 (1985)
35. E. Laviron, AC polograpy and faradaic impedance of strongly adsorbed electroactive species. 2. Theoretical-study of a quasi-reversible reaction in the case of Framkin isotherm. J. Electroanal. Chem. **105**(1), 25–34 (1979)
36. E. Laviron, AC polarography and faradaic impedance of strongly adsorbed electroactive species. 1. Theoretical and experimental-study of quasi-reversible reaction in the case of Langmuir isotherm. J. Electroanal. Chem. **97**(2), 135–149 (1979)
37. A.L. Eckermann, D.J. Feld, J.A. Shaw, T.J. Meade, Electrochemistry of redox-active self-assembled monolayers. Coord. Chem. Rev. **254**(15–16), 1769–1802 (2010)
38. P.R. Bueno, J.J. Davis, G. Mizzon, Capacitance spectroscopy: A versatile approach to resolving the redox density of states and kinetics in redox-active self-assembled monolayers. J. Phys. Chem. C. **116**(30), 8822–8829 (2012)
39. P.R. Bueno, D.A. Miranda, Conceptual density functional theory for electron transfer and transport in mesoscopic systems. Phys. Chem. Chem. Phys. **19**(8), 6184–6195 (2017)
40. S.K. Dey, Y.T. Long, S. Chowdhury, T.C. Sutherland, H.S. Mandal, H.B. Kraatz, Study of electron transfer in ferrocene-labeled collagen-like peptides. Langmuir **23**(12), 6475 (2007)
41. H.S. Mandal, H.B. Kraatz, Electron transfer mechanism in helical peptides. J. Phys. Chem. A Lett. **3**(6), 709–713 (2012)
42. A. Shah, B. Adhikari, S. Martic, A. Munir, S. Shahzad, K. Ahmad, H.B. Kraatz, Electron transfer in peptides. Chem. Soc. Rev. **44** (4), 1015–1027 (2015)
43. X.Y. Xiao, B.Q. Xu, N.J. Tao, Conductance titration of single-peptide molecules. J. Am. Chem. Soc. **126**(17), 5370–5371 (2004)
44. X.L. Li, J. He, J. Hihath, B.Q. Xu, S.M. Lindsay, N.J. Tao, Conductance of single alkanedithiols: conduction mechanism and effect of molecule-electrode contacts. J. Am. Chem. Soc. **128**(6), 2135–2141 (2006)
45. W.Y. Wang, T. Lee, M.A. Reed, Mechanism of electron conduction in self-assembled alkanethiol monolayer devices. Phys. Rev. B **68**(3), (2003)
46. P. Senet, Kohn-Sham orbital formulation of the chemical electronic responses, including the hardness. J. Chem. Phys. **107**(7), 2516–2524 (1997)
47. F.C.B. Fernandes, A. Santos, D.C. Martins, M.S. Goes, P.R. Bueno, Comparing label free electrochemical impedimetric and capacitive biosensing architectures. Biosens. Bioelectron. **57**, 96–102 (2014)

48. F.C.B. Fernandes, M.S. Góes, J.J. Davis, P.R. Bueno, Label free redox capacitive biosensing. Biosens. Bioelectron. **50**, 437–440 (2013)
49. F.C.B. Fernandes, A.V. Patil, P.R. Bueno, J.J. Davis, Optimized diagnostic assays based on redox tagged bioreceptive interfaces. Anal. Chem. **87**(24), 12137–12144 (2015)
50. A.V. Patil, F.C.B. Fernandes, P.R. Bueno, J.J. Davis, Immittance electroanalysis in diagnostics. Anal. Chem. **87**(2), 944–950 (2015)
51. J. Piccoli, A. Santos, N.A. Santos, P.R. Bueno, E.M. Cilli, Peptide in capacitance electroanalysis for diagnostics. J. Pept. Sci. **22**, S24–S24 (2016)
52. J.P. Piccoli, A. Santos, N.A. Santos, E.N. Lorenzon, E.M. Cilli, P.R. Bueno, The self-assembly of redox active peptides: synthesis and electrochemical capacitive behavior. Biopolymers **106**(3), 357–367 (2016)
53. J.P. Piccoli, N.A. Santos, E.N. Lorenzon, A. Santos, F.C.B. Fernandes, P.R. Bueno, E.M.Cilli, Ferrocene-peptides: a new approach for self-assembled monolayers. J. Pept. Sci. **20**, S204–S205 (2014)
54. S. Fletcher, A non-Marcus model for electrostatic fluctuations in long range electron transfer. J. Solid State Electrochem. **11**(7), 965–969 (2007)
55. S. Fletcher, The theory of electron transfer. J. Solid State Electrochem. **14**(5), 705–739 (2010)
56. R.A. Marcus, N. Sutin, Electron transfers in chemistry and biology. Biochim. Biophys. Acta **811**(3), 265–322 (1985)
57. J. Lehr, J.R. Weeks, A. Santos, G.T. Feliciano, M.I.G. Nicholson, J.J. Davis, P.R. Bueno. Mapping the ionic fingerprints of molecular monolayers. Phys. Chem. Chem. Phys. (2017)
58. P.R. Bueno, G.T. Feliciano, J.J. Davis, Capacitance spectroscopy and density functional theory. Phys. Chem. Chem. Phys. **17**, 9375–9382 (2015)
59. J. Lehr, F.C.B. Fernandes, P.R. Bueno, J.J. Davis, Label-free capacitive diagnostics: exploiting local redox probe state occupancy. Anal. Chem. **86**(5), 2559–2564 (2014)
60. A. Santos, F.C. Carvalho, M.C. Roque-Barreira, P.R. Bueno, Impedance-derived electrochemical capacitance spectroscopy for the evaluation of lectin-glycoprotein binding affinity. Biosens. Bioelectron. **62**, 102–105 (2014)
61. A. Santos, P.R. Bueno, Glycoprotein assay based on the optimized immittance signal of a redox tagged and lectin-based receptive interface. Biosens. Bioelectron. **83**, 368–378 (2016)
62. S.M. Marques, A. Santos, L.M. Goncalves, J.C. Sousa, P.R. Bueno, Sensitive label-free electron chemical capacitive signal transduction for D-dimer electroanalysis. Electrochim. Acta **182**, 946–952 (2015)
63. J.S. Hwang, K.J. Kong, D. Ahn, G.S. Lee, D.J. Ahn, S.W. Hwang, Electrical transport through 60 base pairs of poly(dG)-poly(dC) DNA molecules. Appl. Phys. Lett. **81**(6), 1134–1136 (2002)
64. D. Porath, A. Bezryadin, S. de Vries, C. Dekker, Direct measurement of electrical transport through DNA molecules. Nature **403**(6770), 635–638 (2000)
65. N.J. Tao, Electron transport in molecular junctions. Nat. Nanotechnol. **1**(3), 173–181 (2006)
66. A.K. Mahapatro, K.J. Jeong, G.U. Lee, D.B. Janes, Sequence specific electronic conduction through polyion-stabilized double-stranded DNA in nanoscale break junctions. Nanotechnology **18**(19), 195202 (2007)
67. E. Meggers, M.E. Michel-Beyerle, B. Giese, Sequence dependent long range hole transport in DNA. J. Am. Chem. Soc. **120**(49), 12950–12955 (1998)
68. K.H. Yoo, D.H. Ha, J.O. Lee, J.W. Park, J. Kim, J.J. Kim, H.Y. Lee, T. Kawai, H.Y. Choi, Electrical conduction through Poly(dA)-Poly(dT) and Poly(dG)-Poly(dC) DNA molecules. Phys. Rev. Lett. **87**(19), 198102 (2001)
69. L.T. Cai, H. Tabata, T. Kawai, Self-assembled DNA networks and their electrical conductivity. Appl. Phys. Lett. **77**(19), 3105–3106 (2000)
70. W.C. Ribeiro, L.M. Goncalves, S. Liebana, M.I. Pividori, P.R. Bueno, Molecular conductance of double-stranded DNA evaluated by electrochemical capacitance spectroscopy. Nanoscale **8**(16), 8931–8938 (2016)
71. E. Wierzbinski, R. Venkatramani, K.L. Davis, S. Bezer, J. Kong, Y. Xing, E. Borguet, C. Achim, D.N. Beratan, D.H. Waldeck, The single-molecule conductance and electrochemical electron-transfer rate are related by a power law. Acs Nano **7**(6), 5391–5401 (2013)

72. H.B. Heersche, P. Jarillo-Herrero, J.B. Oostinga, L.M.K. Vandersypen, A.F. Morpurgo, Bipolar supercurrent in graphene. Nature **446** (7131), 56–59 (2007)

73. A.V. Patil, F.B. Fernandes, P.R. Bueno, J.J. Davis, Graphene-based protein biomarker detection. Bioanalysis **7**(6), 725–742 (2015)

74. X. Zhang, B.R.S. Rajaraman, H. Liu, S. Ramakrishna, Graphene's potential in materials science and engineering. RSC Advances **4**(55), 28987–29011 (2014)

75. F. Sharifi, S. Ghobadian, F.R. Cavalcanti, N. Hashemi, Paper-based devices for energy applications. Renew. Sustain. Energy Rev. **52**, 1453–1472 (2015)

76. J. Xia, F. Chen, J. Li, N. Tao, Measurement of the quantum capacitance of graphene. Nat. Nanotechnol. **4**(8), 505–509 (2009)

77. F.A. Gutierrez, F.C.B. Fernandes, G.A. Rivas, P.R. Bueno, Mesoscopic behaviour of multi-layered graphene: the meaning of supercapacitance revisited. Phys. Chem. Chem. Phys. **19**(9), 6792–6806 (2017)

78. P.R. Bueno, J.J. Davis, Measuring quantum capacitance in energetically addressable molecular layers. Anal. Chem. **86**, 1337–1341 (2014)

79. Y.Q. Xue, M.A. Ratner, Theoretical principles of single-molecule electronics: A chemical and mesoscopic view. Int. J. Quantum Chem. **102**(5), 911–924 (2005)

80. M. Brandbyge, J.L. Mozos, P. Ordejon, J. Taylor, K. Stokbro. Density-functional method for nonequilibrium electron transport. Phys. Rev. B. **65**(16) (2002)

81. M. Bruchez, M. Moronne, P. Gin, S. Weiss, A.P. Alivisatos, Semiconductor nanocrystals as fluorescent biological labels. Science **281**(5385), 2013–2016 (1998)

82. W.C.W. Chan, S.M. Nie. Quantum dot bioconjugates for ultrasensitive nonisotopic detection. Science **281**(5385), 2016–2018 (1998)

83. X. Michalet, F.F. Pinaud, L.A. Bentolila, J.M. Tsay, S. Doose, J.J. Li, G. Sundaresan, A.M. Wu, S.S. Gambhir, S. Weiss, Quantum dots for live cells, in vivo imaging, and diagnostics. Science **307**(5709), 538–544 (2005)

84. A.P. Alivisatos. Semiconductor clusters, nanocrystals, and quantum dots. Science **271**(5251), 933–937 (1996)

85. C.W.J. Beenakker, H. Vanhouten, Quantum transport in semiconductor nanostructures. Solid State Phys. **44**, 1–228 (1991)

86. T. Mokari, E. Rothenberg, I. Popov, R. Costi, U. Banin, Selective growth of metal tips onto semiconductor quantum rods and tetrapods. Science **304**(5678), 1787–1790 (2004)

87. F.F. Hudari, G.G. Bessegato, F.C.B. Fernandes, M.V.B. Zanoni, P.R. Bueno, Reagentless detection of low-molecular-weight triamterene using self-doped TiO_2 Nanotubes. Anal. Chem. (2018)

88. J.R. Lakowicz, Radiative decay engineering: biophysical and biomedical applications. Anal. Biochem. **298**(1), 1–24 (2001)

Final Remarks

Foremost Conclusions

Electrochemical capacitance, $C_{\bar{\mu}}$, as supported by first-principle quantum mechanics, was herein overviewed in the context of its impact on nanoscale electrochemistry and molecular electronics. The associated energy $\left(\propto 1/C_{\bar{\mu}}\right)$ defines the electron transfer rate as $k = G/C_{\bar{\mu}}$, in which the electron conductance G is the proportional term. Accordingly, $C_{\bar{\mu}}$ can be defined directly through an electron transfer rate (k) and Landauer conductance (G), such that $C_{\bar{\mu}} = G/k$. Based on this definition of $C_{\bar{\mu}}$, we therefore demonstrated that double-layer and pseudocapacitances—the latter empirically qualified but poorly understood—are fully incorporated in the definition of $C_{\bar{\mu}}$. In other words, both double-layer and pseudocapacitances were shown to be merely particular estimations of $C_{\bar{\mu}}$, depending solely on the type of electric field screening taking place, namely Debye (non-faradaic) or Thomas–Fermi (faradaic) screening.

Of particular importance is the finding that molecular electroactive films self-assembled on electrodes serve as an exemplary ensemble of quantum point contacts electronically coupled to microscopic electrodes. This finding not only establishes an important mesoscopic statistical behaviour but also conceptually demonstrates that the relationship between G, $C_{\bar{\mu}}$ and k holds beyond atomistic viewpoint of a single particle. The timescale of charging processes governed by $C_{\bar{\mu}} = G/k$ was demonstrated to be a simple case of a quantum RC circuit, whose electron dynamics is attenuated by the presence of an electrolyte. Because of the dependence on an electrolyte, as a matter of physical chemistry definition, this particular circuit was presented as a quantum electrochemical RC circuit.

Therefore, the only difference between nanoelectronics and nano-electrochemistry is the presence of an electrolyte in the latter, within a shielded electrical field, which governs the equilibrium electrostatic potential of capacitors. In general, it can thus be assumed that the physics of nanoscale electronics and electrochemistry is

P. R. Bueno, *Nanoscale Electrochemistry of Molecular Contacts*, SpringerBriefs in Applied Sciences and Technology, https://doi.org/10.1007/978-3-319-90487-0

fundamentally the same and has been shown to be governed by conjoint quantum mechanical principles.

In summary, based on the universal definition of electrochemical capacitance, $C_{\bar{\mu}} = G/k$, we demonstrated that nanoscale electrochemistry matches molecular electronics in a straightforward manner. Implicitly, it can be noted that the deeply rooted definition of charge transfer resistance (R_{ct}) is replaced by its quantum mechanics equivalent, $G = 1/R_{ct}$. Lastly, the applicability of the concepts was illustrated in molecular diagnostics and served to demonstrate the origin of the supercapacitive behaviour of graphene compounds, in which the perceptions and elucidative interpretation of $C_{\bar{\mu}} = G/k$ for the particular systems were developed on a case-by-case basis. The impacts that the $C_{\bar{\mu}} = k/G$ relationship may have and how it will succeed in integrating electron transfer and mesoscopic transport theories remains largely unknown, with a wealth of unexplored information yet to be revealed.